Energy and Transience

-The Journey After Death-

Drake Graeve

Energy and Transience -The Journey After Death

Copyright © 2024 Drake Graeve

Alle Rechte vorbehalten.

ISBN: 9798304248761

*In recognition of those who courageously explore
the uncharted territories of life after death.*

*With rigorous scholarship and open minds,
they illuminate the complexities
of human existence.*

Inhaltsverzeichnis

Preface i
Introduction 3
    Understanding the Transition: Energy and Death 3
    The Role of Energy in Life and Death 6
    Purpose of the Book 9
Scope of This Book 13
    The Importance of Understanding Energy Transformation 14
Structure of the Book 17
Chapter 1: The Human Body as an Energy System 19
    1.1. Chemical Composition of the Human Body 21
1.1.1. Fats 21
1.1.2. Proteins 23
1.1.3. Carbohydrates 24
    1.2. Energy metabolism in the body 27
1.2.1. Catabolism 27
1.2.2. Anabolism 29
    1.3. Energy Balance and Physiological Functions 33
1.3.1. Basal Metabolic Rate (BMR) 33
1.3.2. Physical Activity 35
1.3.3. Thermogenesis 36
Chapter 2: The Final Moments of Life 39
    2.1. Physiological Changes Immediately Before Death 41
2.1.1. The Process of Dying 41
2.1.2. Neurological Changes 44
    2.2. Biochemical Processes at the Time of Death 47
2.2.1. Energy Depletion and Metabolic Changes 47
2.2.2. Onset of Autolysis 49
2.2.3. Role of Enzymes 50
Chapter 3: At the Moment of Death 53
    3.1. Defining Death 55
3.1.1. Clinical Death 57
3.1.2. Biological Death 58
    3.2. Immediate Pphysiological Reactions at the

Moment of Death 63
   3.2.1. Cardiovascular System 64
   3.2.2. Respiration 66
   3.2.3. Physical Changes 67
      3.3. Chemical and biochemical changes 71
   3.3.1. Loss of Energy...71
   3.3.2. Acid-Base Balance 73
   3.3.3. Enzymatic Decomposition 74
      3.4. The transition to decomposition 77
   3.4.1. Early Decomposition: Rigor Mortis...78
   3.4.2. Postmortem Lividity (Livor Mortis)...80
   3.4.3. Microbial Decomposition 82
   3.4.4. Odor Formation 84
Chapter 4: The first hours after death 87
      4.1. Early physiological changes 89
   4.1.1. Temperature Changes...89
   4.1.2. Blood Clotting...91
   4.1.3. Muscle Relaxation and Rigor Mortis 93
      4.2. Biochemical processes and enzymatic activities 97
   4.2.1. Enzymatic Decomposition 97
      4.3. Microbiological changes 103
   4.3.1. Microbial Invasion..103
   4.3.2. Gas Formation 106
      4.4. Other physiological changes 109
   4.4.1. Skin Changes..109
   4.4.2. Odor Formation 111
Chapter 5: The process of decomposition..113
      5.1. Introduction of decompositionin..115
   5.1.1. Definition and Stages of Decomposition 116
      5.2. Detailed examination of the decomposition stages 119
   5.2.1. The Fresh Stage 119
   5.2.2. The Bloating Stage...121
   5.2.3. The Decay Stage 122
   5.2.4. The Dry Stage 124
      5.3. Factors influencing Decomposition 127
   5.3.1. Temperature 127

- 5.3.2. Moisture 128
- 5.3.3. Environmental Conditions 129
  - 5.4. Microbiological Aspects of Decomposition 131
- 5.4.1. Microbes and Bacteria 131
- 5.4.2. Microbiomes and Decomposition 133

Chapter 6: Long-term energy Transformations 137
  - 6.1. Energy Transformations after Decomposition 139
- 6.1.1. Conversion of Organic Molecules 139
- 6.1.2. Mineralization 141
  - 6.2. Impact on the Ecological System 145
- 6.2.1. Nutrient Cycles 145
- 6.2.2. Energy Flows in Ecosystems 147
  - 6.3. Long-Term Effects on Soil 151
- 6.3.1. Soil Improvement 151
- 6.3.2. Soil Chemical Changes 153
  - 6.4. Long-Term Energy Transformations in Nature 157
- 6.4.1. The Formation and Role of Humus 157
- 6.4.2. Impact on Biological Activity 160

Chapter 7: Ecological and physical perspectives 163
  - 7.1. Ecological Perspectives 167
- 7.1.1. The Role of Decomposition in the Nutrient Cycle ...167
- 7.1.2. Effects on Soil Biology 169
- 7.1.3. Effects on Plant Growth 173
- 7.1.4. Ecological Dynamics 176
  - 7.2. Physical Perspectives 179
- 7.2.1. Temperature and Moisture Conditions 179
- 7.2.2. Physical Weathering Processes 182
- 7.2.3. Energy Transfer and Loss 184
- 7.2.4. Effects on Soil Physics 187
  - 7.3. Practical Applications and Research 191
- 7.3.1. Agricultural Applications 191
- 7.3.2. Environmental Research 193
  - 7.4. Summary 197

Chapter 8: Philosophical and cultural considerations 201
  - 8.1. Philosophical reflections on Death 207
- 8.1.1. Philosophical Views on Death 207

- 8.1.2. The Concept of Immortality 209
  - 8.2. Cultural considerations and rituals 213
- 8.2.1. Funeral Rites and Traditions 213
- 8.2.2. Cultural Perspectives on Decomposition 216
- 8.2.3. Symbolism of Death and Decomposition 218
  - 8.3. Decomposition's Role in Philosophy and Ethics 223
- 8.3.1. Environmental Ethics and Sustainability 223
- 8.3.2. The Role of Decomposition in Cultural Identity 226
  - 8.4. Summary 229
- Concluding Thoughts 237
- Appendix 239
  - A.1 Glossary of Terms 239
  - A.2 References 241
  - A.3 Methodology 242
  - A.4 Sources for Further Information 242
  - A.5 Acknowledgments 243
  - A.6 Liability 243
  - A.7 Contact Information 243

Preface

In human existence, there are two unavoidable constants: energy and transience. These forces shape our lives and our understanding of the world. But what happens when our bodies perish and life comes to an end? What becomes of the energy that carried us through life? These questions have preoccupied humanity for centuries and are the core focus of this book.

In *Energy and Transience – The Journey After Death*, I invite you to embark on a journey beyond known reality. Here, death is not viewed as a definitive end but as a transition. While the finiteness of our physical existence is inevitable, the question of what follows remains one of the most intriguing challenges to the human mind.

This book is not a scientific treatise or a dogmatic text on life after death. It is a reflection on the immortal energy that connects us and how this energy might continue to exist after physical death. It explores various perspectives and traditions dealing with life after death and encourages readers to question and expand their own viewpoints.

We live in a world shaped by rationality and science. Yet, at the boundaries of our existence, we encounter phenomena that cannot always be scientifically explained. The myriad stories and experiences from different cultures and epochs reveal our profound need to understand and come to terms with death.

In the chapters ahead, we will explore questions that define this timeless quest: What remains of us when our bodies die? How can energy, which cannot be destroyed, persist? Is there consciousness after death, and what might it look like? We will delve into religious, philosophical, and spiritual traditions and incorporate new insights from quantum physics and near-death research.

This book does not aim to provide definitive answers but rather to spark reflection. It seeks to encourage engagement with the profound questions of life and death, offering space for speculative thinking and creative exploration. You may not find absolute answers, but perhaps new insights and perspectives that help you see death not as an end but as part of a larger process.

I invite you to read this book with an open heart and a curious mind. May it inspire you to look beyond what we perceive as reality and entertain the thought that life, in whatever form, might never truly end.

Introduction

Understanding the Transition: Energy and Death

Death is a universal experience that has fascinated humanity throughout history. From the earliest days of civilization, humans have sought to uncover the mysteries surrounding this pivotal moment of existence. Ancient cave paintings, the myths and legends of indigenous peoples, and the sacred texts of major religions reflect humanity's efforts to understand death and what follows.

These age-old questions have inspired countless cultures and philosophies to explore the meaning and implications of death, seeking answers that touch every one of us. What happens when life ends? What becomes of the essence of who we are? These profound questions have given rise to diverse interpretations, rituals, and beliefs that vary across cultures and religions.

In recent centuries, science has begun to examine these ancient questions through the lens of rationality and empiricism. Advances in biology, chemistry, and physics have helped us view death not only as an end but also as a transition. The energetic aspects of this transition, in particular, offer a unique perspective.

Death is not simply the conclusion of a biological process; it is the beginning of a remarkable transformation. From the moment of death to the eventual recycling of biological materials in the environment, this profound journey of energy transformation unfolds. These processes are not only fascinating but also fundamental to understanding the role of death in the cycle of life.

When contemplating death, many envision a sudden, final silence. Yet, in reality, death marks the start of an incredibly dynamic and complex chain of events. The energy that sustains our bodies throughout life does not simply vanish with death; it continues to exist in new and transformed forms.

This transformation occurs through biological decomposition processes, during which microorganisms break down organic materials, releasing stored energy. This released energy enters the food chain, supporting the growth of new life forms. Understanding these processes allows us to recognize the deeper connection between life and death, impermanence and renewal.

The human body, composed of countless cells, holds an immense amount of energy. This energy is stored in chemical bonds and is continuously utilized and renewed during life. At death, this constant renewal ceases, and another process begins: biological decomposition. Microorganisms within and around our bodies take over, breaking down organic materials. This process is an essential part of the natural cycle, whereby the energy that once animated our bodies is released and repurposed.

The science behind this biological decomposition is as fascinating as it is complex. Various bacteria and fungi play crucial roles in this process, breaking down organic materials into simpler molecules that can be absorbed by plants and other organisms. In this way, the energy stored in the body is returned to the environment, contributing to the food chain. Plants use these nutrients to grow and store energy through photosynthesis, and animals consuming these plants utilize this energy, continuing the cycle.

This book aims to shed light on and explore these processes in detail, examining how the energy stored in the human body transforms after death. We will delve into the science of biological decomposition, the role of microorganisms, and the conversion of organic materials into new forms of life. Beyond the physical and chemical aspects, we will also consider the biological and ecological consequences. Additionally, we will explore how these insights might influence our understanding of life and death, as well as our perceptions of impermanence.

Death is more than an end—it is a transition, a transformation with the potential to profoundly alter our understanding of life and existence. By examining the energetic processes that occur after death, we can better understand biological mechanisms and the deeper significance of death as an integral part of the life cycle.

This book invites you to join us on this journey, exploring the mysteries of death and gaining new perspectives on the energy and transience of life. By investigating these energetic transformations, we aim to bridge the gap between scientific explanations and the philosophical and spiritual aspects of death, creating a more comprehensive picture of this fascinating subject.

The process of transformation, from the moment of death to the ultimate recycling of biological materials in the environment, is a profound journey of energy conversion. These transformations are not only captivating but also crucial for understanding the role death plays in the cycle of life.

The energy that sustains our bodies during life does not simply disappear at death; it persists in various forms, undergoing transformation. Through biological decomposition, microorganisms break down organic materials, releasing stored energy into the environment. This energy supports the food chain and fosters the growth of new life forms. Understanding these processes can help us appreciate the intricate connection between life and death, between transience and renewal.

Death is a phenomenon that touches all living beings, yet it remains one of the most misunderstood and feared events in human life. The idea that death is not an end but a transition can fundamentally change how we understand life and death. By closely examining the energetic processes that follow death, we can gain not only a better understanding of biological mechanisms but also a deeper appreciation of death as an integral part of the life cycle.

The connection between life and death, impermanence and renewal, is profound and complex. Understanding this connection can help us live more mindfully and cherish the natural world around us.

This book is an invitation to dive deeper into these fascinating processes and uncover the remarkable mechanisms that shape life after death. We hope that by reading this book, you will gain a deeper understanding of the energetic transformations that occur after death and a new appreciation for the beauty and complexity of life and death.

The Role of Energy in Life and Death

Energy is fundamental to life and plays a central role in all biological processes. Every cellular function, from muscle contraction to cognitive processes, relies on the transformation and utilization of energy. Our body is a complex system that constantly absorbs, converts, and utilizes energy to sustain its diverse functions. Throughout life, our body manages energy through intricate biochemical pathways, maintaining a balance between intake and consumption. This energy is derived from the food we eat and stored in accessible forms, such as adenosine triphosphate (ATP), the universal energy source for cells.

The role of energy in life is profound, spanning every aspect of biological function. Every single cellular activity—be it muscle contraction essential for movement or cognitive processes that enable thinking, remembering, and learning—depends on the continuous transformation and utilization of energy. Indeed, the human body is an extraordinarily complex and finely tuned system, constantly drawing energy from various sources, converting it into usable forms, and efficiently utilizing it to perform a wide range of vital functions that sustain and enhance life. Over the course of life, our body meticulously regulates this energy through a series of sophisticated biochemical mechanisms, ensuring a balance between the energy derived from food and its consumption for bodily needs. This energy is stored in forms readily accessible for cellular activities, with ATP serving as the universal fuel for all cells.

Throughout our lives, the body is designed to use energy efficiently. Energy sustains bodily functions, promotes growth and repair, and enables interaction with the environment. This process requires continuous adaptation and fine-tuning to respond to changing conditions and demands. However, this dynamic balance undergoes a dramatic shift when life comes to an end. Death marks the cessation of these carefully orchestrated processes and initiates a series of transformations, redistributing stored energy into various forms.

During life, energy fuels a wide range of bodily functions, from basic processes like heartbeat and respiration to more complex activities such as tissue growth, repair, and interaction with the environment. This process requires constant adjustments to ensure the body can respond to changing conditions and demands. The body is perpetually engaged in managing, storing, and utilizing energy to sustain all essential functions. However, when life ends, this dynamic equilibrium is fundamentally altered. Death signifies the conclusion of these intricately coordinated processes and triggers a cascade of transformations in which stored energy is converted into other forms.

After death, a new phase of energy flow begins. The energy stored within the body does not simply vanish; instead, it is redistributed into other forms and systems. The biological decomposition process commences, with microorganisms such as bacteria and fungi playing a crucial role. These organisms break down the body's organic compounds, releasing energy into the environment. This energy is then utilized by other organisms, contributing to the cycle of life. The breakdown of organic material is thus a vital aspect of the energy transformation process that sustains life on Earth.

This phase of energy flow following death is both fascinating and essential. The energy stored in the body during life does not disappear; it is transformed and integrated into the broader natural world. The decomposition process, driven by microorganisms like bacteria and fungi, breaks down the body's organic matter, releasing energy that reenters the environment. This released energy is absorbed by other organisms and becomes part of the cycle of life. The decomposition of organic material is a critical component of the energy transformation that perpetually drives and sustains life on Earth.

Understanding these transformations provides not only insights into the biological processes involved but also reveals broader ecological and philosophical implications. The conversion of energy after death highlights the close interconnection between life and death, illustrating how death is an integral part of the natural cycle. This perspective allows us to view death not merely as an end but as a continuation of life in a transformed state. By examining the role of energy in life and death, we gain a deeper understanding of our existence and environment, recognizing that the energy that once animated us remains a part of the larger ecosystem.

This understanding offers valuable insights into the biological processes during death while also deepening our awareness of the broader ecological and philosophical ramifications. The transformation of energy following death underscores the intimate connection between life and death, revealing that death is, in fact, a fundamental part of the natural cycle. This perspective enables us to see death not as an absolute end but as a continuation of life in a new, transformed form. Exploring the role of energy in life and death cultivates a comprehensive understanding of our existence and the intricate relationships within our environment. We come to realize that the energy that sustained and propelled us during life does not vanish but continues to be an active part of the vast, interconnected ecosystem.

Purpose of the Book

This book examines the intricate processes involved in the transformation of energy from the moment of death through decomposition to its integration into ecosystems. It aims to provide a comprehensive overview of how energy transitions from its stored form in the human body to various environmental states, contributing to the cycle of life and matter. By exploring the different stages and mechanisms of this transformation in detail, the book seeks to foster a deep understanding of how energy continues to influence life after death. Through an in-depth examination of each step in this transitional journey, readers are invited into the fascinating world of energy transformation, illustrating the inseparable connection between life and death.

The journey of energy does not end with death; rather, it marks the beginning of a remarkable transition where biological, chemical, and physical processes converge to recycle and redistribute energy. This book illuminates these complex mechanisms, emphasizing their essential role in sustaining life on Earth. Energy transformation begins immediately after death and progresses through various stages, including decomposition by microorganisms, biochemical release of energy carriers, and the eventual integration of this energy into surrounding ecosystems.

Detailed analyses will provide readers with insights into the roles of microorganisms in decomposition, the biochemical pathways of energy release, and the ultimate reintegration of this energy into ecosystems. Furthermore, the ecological impacts of these processes will be explored, demonstrating how released energy supports new life forms and maintains ecological balance.

The scope of this book extends beyond biological processes to explore the philosophical and ecological implications of energy transformation. By understanding how energy is recycled in nature, we can better appreciate the interconnectedness of all living things and the delicate balance that sustains life. This book also examines how these processes reflect broader ecological principles and underscore the importance of preserving this balance in the face of environmental change. It highlights how the natural processes of energy conversion are vital not only for biodiversity but also for the functionality of entire ecosystems.

From a philosophical perspective, the book underscores the significance of death within the context of the life cycle and its role in continual renewal. Readers are encouraged to recognize and understand the deeper meaning of death—not as an end but as a transition to a new state of being where the energy that once powered a living organism remains active within the cycle of nature, supporting new forms of life.

Additionally, this book aims to bridge the gap between scientific understanding and philosophical reflection on life and death. It encourages readers to contemplate the continuity of life through energy transformation and to reconsider the perception of death as finality. By exploring these themes, the book hopes to foster a deeper appreciation for the cycles of life and the role of energy in our existence. Readers are invited to reflect on the awe-inspiring processes that unfold after the death of a living being and to see death not as a definitive end but as a transition to new forms of life and energy. By examining these energetic transformations, we can develop a new perspective on our existence and the ongoing connections between life and death.

This understanding can help us view death not as a loss but as an essential part of the natural cycle that continuously brings forth new life. Through a multidisciplinary approach, this book offers a thorough exploration of energy transformation after death, shedding light on the scientific, ecological, and philosophical aspects of this process. Readers are encouraged to consider how the energy that once fueled our lives continues to affect the world around us, revealing the enduring links between life, death, and the natural world.

Scientific insights are combined with philosophical reflections to provide a holistic view of energy conversion. The chapters of this book delve into the role of microorganisms in decomposition, the chemical processes involved, and the ways in which released energy reenters the natural cycle. This offers a comprehensive perspective on the significance of death within the context of life, showing how intricately woven and interdependent the processes of living and dying truly are.

Through this comprehensive and interdisciplinary approach, the book aspires to create a profound understanding of the energetic transformations that shape both our lives and our environment.

Energy and Transience -The Journey After Death

Scope of This Book

The scope of this book is comprehensive, encompassing multiple dimensions of the processes and implications of energy transformation after death. Each facet is explored in detail, providing a thorough understanding of the interconnected roles of biology, chemistry, ecology, and philosophy.

Biological Processes
This section delves into the physiological changes that occur at the moment of death, the early stages of decomposition, and the pivotal role microorganisms play in breaking down the body. By examining these processes step by step, we uncover the intricate mechanisms that drive the transformation of organic matter and prepare it for reintegration into the environment.

Energetic Transformations
Here, we analyze how the chemical energy stored in the human body is converted into heat, gases, and other byproducts during decomposition. This transformation is not only a chemical marvel but also a key element in the cycle of life, illustrating how energy persists and changes form even after death.

Ecological Impacts
The ecological significance of decomposition is explored in depth, highlighting how the materials and energy released during this process contribute to nutrient cycles and energy flows within ecosystems. These contributions underscore the essential role decomposition plays in sustaining life and maintaining the balance of natural systems.

Philosophical and Cultural Perspective
Beyond the scientific aspects, the book delves into cultural perceptions of death and their connection to energy and matter. This section reflects on the diverse ways in which

societies interpret death, offering insights into how these views shape our understanding of life's continuity and interconnectedness.

By integrating these themes, the book not only explains the scientific phenomena behind energy transformation but also offers a broader perspective on the relationship between life, death, and the natural world. This multidisciplinary approach invites readers to appreciate the complexity and beauty of these processes while encouraging deeper contemplation of their significance in our lives and environments.

The Importance of Understanding Energy Transformation

Understanding the transformation of energy after death has both practical and theoretical implications. On a practical level, it deepens our knowledge of ecological processes and the critical role decomposition plays in nutrient cycles. The breakdown of organic material releases nutrients that can be absorbed by plants and other organisms, preserving soil fertility and enhancing the productivity of ecosystems. This understanding is also valuable in fields such as agriculture, forestry, and conservation, where optimizing natural cycles is essential for maintaining healthy and sustainable environments.

In agriculture, for example, knowledge of decomposition and energy transformation can be used to improve composting processes. By encouraging the efficient breakdown of organic waste, farmers can produce high-quality fertilizers that enhance soil quality and yield better crops. In forestry, understanding decomposition helps manage natural processes in forests, where dead trees and plants provide nutrients for new growth, contributing to forest regeneration. In conservation, this knowledge is invaluable for preserving and enhancing habitats. Targeted interventions can support natural decomposition processes, fostering biodiversity and strengthening ecosystem health.

Theoretically, understanding energy transformation after death offers a deeper appreciation for the continuity of energy through life and death. This continuity links biological processes to broader environmental cycles, demonstrating that energy is never lost but merely changes form. This perspective highlights the elegance and complexity of the natural world and reveals the profound interconnectedness of all living forms. By exploring the energetic transformations that occur after death, we gain a clearer understanding of the relationships between biotic (living) and abiotic (non-living) components of ecosystems.

This understanding unveils the intricate interactions between organisms and their environments. Biotic components, such as living organisms, and abiotic components, like soil, water, and air, are inseparably connected through energy transformations. When an organism dies, the decomposition process releases energy and nutrients that can be utilized by other organisms, creating a continuous cycle that sustains life on Earth. This deeper appreciation of natural processes can inspire us to interact more responsibly with our environment, safeguarding the delicate balance of ecosystems.

Additionally, this knowledge provides valuable perspectives on the connection between life and death, contributing to philosophical and cultural reflections on mortality and the natural world. By observing the cycle of energy transformation, we can view death not as an end but as an integral part of the life cycle. This perspective enriches our understanding of mortality, helping us accept death as a natural and necessary process that supports life in its myriad forms.

Death, therefore, is not the end of a life but the beginning of a new chapter in nature's cycle. The energy stored in an organism during its lifetime is released after death, fueling the emergence of new life. This understanding helps frame death not as a loss but as a natural and continuous process. Many cultures and philosophies already embrace this perspective, and scientific insights into energy transformation can further reinforce it. It invites us to recognize the beauty and harmony of nature's cycles, prompting us to reconsider our views on mortality and life.

Through a detailed exploration of these processes, we gain insights into the fundamental principles governing life and death, enriching our understanding of both. These insights can also influence our cultural and spiritual perspectives, encouraging contemplation of the significance of life, death, and the role of energy in the universe. Ultimately, understanding energy transformation contributes to a deeper knowledge of the world we inhabit, allowing us to appreciate the complexity and beauty of the natural processes that sustain life.

This deeper understanding can inspire us to rethink our lifestyles and our relationship with nature. It encourages more sustainable and eco-friendly decisions that respect and support natural cycles. Awareness of the continuous transformation of energy fosters a greater appreciation for the resources we have and underscores the importance of protecting and conserving them. In the end, understanding energy transformation after death highlights the profound interconnectedness of all living beings and underscores the significance of the natural processes that make life on Earth possible.

Structure of the Book

This book is structured to guide readers through the various phases of energy transformation and decomposition, providing a comprehensive understanding of the processes and their implications:

Chapter 1: The Human Body as an Energy System
Introduces the fundamental principles of energy storage and metabolism in the human body, laying the groundwork for understanding energy's role in life.

Chapter 2: The Final Moments of Life
Describes the physiological and energetic changes that occur as life transitions to death, offering insight into the processes leading to the cessation of biological functions.

Chapter 3: At the Moment of Death
Explores the immediate changes following death, focusing on the first stages of energy transformation and the shifts that mark the start of the decomposition process.

Chapter 4: The First Hours After Death
Details the early phases of decomposition and the associated energy patterns, highlighting the role of microorganisms and initial biophysical changes.

Chapter 5: The Decomposition Process
Examines the biochemical and ecological aspects of decomposition, uncovering the intricate processes that recycle organic matter back into the environment.

Chapter 6: Long-Term Energy Metamorphoses
Discusses how decomposed materials integrate into ecological cycles, emphasizing the continuity of energy and its vital role in sustaining ecosystems.

Chapter 7: Ecological and Physical Perspectives
Offers insights into the broader environmental impacts of

decomposition, connecting the energy transformation process to global ecological dynamics.

Chapter 8: Philosophical and Cultural Reflections
Explores cultural perceptions of death and energy transformation, examining how societies interpret these processes and their significance.

Conclusion
Summarizes the key findings and emphasizes their broader implications for understanding life, death, and the interconnectedness of all living things.

Closing Thoughts
Death is a transition marked by the transformation of stored energy into new forms. The processes involved illustrate the profound connection between life and death within the natural cycle. By understanding energy transformation and the role of decomposition, we gain valuable insights into the continuity of life and the interdependence of all living things.

As we embark on this exploration, readers are invited to engage with both the scientific details and the broader philosophical and ecological implications of these processes. Through this journey, we aim to foster a deeper appreciation for the beauty and complexity of life, death, and the natural cycles that sustain our existence.

Chapter 1:
The Human Body as an Energy System

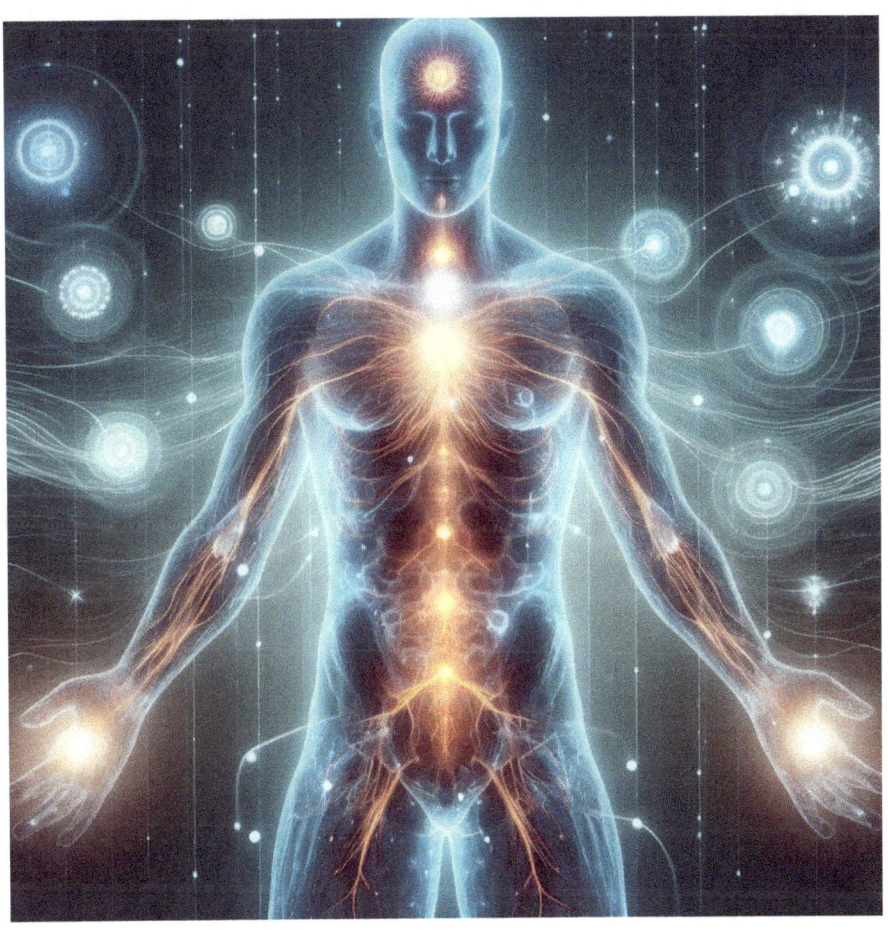

Energy and Transience -The Journey After Death

1.1. Chemical Composition of the Human Body

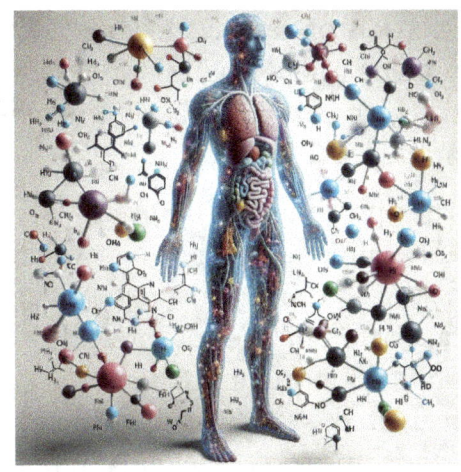

The human body is composed of a diverse array of chemical compounds that serve as vital energy sources. The primary components responsible for energy production and storage include fats, proteins, and carbohydrates. These macronutrients are essential for maintaining life functions and regulating energy balance. Their unique properties and functions enable the body to operate efficiently and adapt to varying demands. Through the continuous transformation and exchange of these chemical compounds, the body supports a wide range of activities and processes crucial for survival and health.

1.1.1. Fats

Fats are the primary source of long-term energy, providing approximately 9 kcal per gram. Stored predominantly as triglycerides, fats are located in adipose tissues such as subcutaneous and visceral fat. These tissues not only serve as energy reserves but also function as insulators, protecting the body from cold and cushioning against mechanical shocks.

Triglycerides

Triglycerides consist of three fatty acids bound to a glycerol molecule, forming an efficient long-term energy storage system. During periods of energy demand, such as physical exertion or fasting, triglycerides are broken down through a process called lipolysis, releasing fatty acids and glycerol into the bloodstream. These are subsequently oxidized in cells to produce ATP.

The process of lipolysis is regulated by hormones like adrenaline and glucagon, which are released during stress or fasting. These hormones activate specific enzymes that facilitate triglyceride breakdown, ensuring the availability of fatty acids for mitochondrial energy production. This tightly regulated system exemplifies the body's ability to mobilize energy reserves efficiently in response to varying needs.

Phospholipids

Phospholipids are key structural components of cell membranes, contributing to their stability and flexibility. While less central as an energy source, they are crucial for cellular structure and function. Phospholipids form a bilayer that acts as a barrier, controlling the exchange of substances between the cell interior and its external environment.

In addition to their structural role, phospholipids participate in cellular communication and signaling processes. They transport signaling molecules in and out of cells, enabling responses to internal and external stimuli. This dual role highlights their importance in maintaining both the physical integrity and dynamic functionality of cells.

Steroids

Cholesterol, a significant steroid, is embedded in cell membranes and serves as a precursor for the synthesis of steroid hormones such as testosterone and estrogen. These hormones regulate critical physiological processes, including growth, metabolism, and reproduction.

Synthesized in the liver and obtained through diet, cholesterol also contributes to the stability and fluidity of cell membranes. It plays a role in forming lipid rafts—specialized microdomains within membranes that are essential for signal transduction. This multifaceted functionality makes cholesterol indispensable for both structural integrity and cellular communication.

1.1.2. Proteins

Proteins are macromolecules composed of amino acids, providing approximately 4 kcal per gram. They are fundamental to the structure, function, and regulation of the body. Proteins participate in nearly all biological processes, fulfilling diverse roles that range from structural support to the facilitation of biochemical reactions.

Structural Proteins

Structural proteins like collagen and elastin are key components of connective tissues and skin. Collagen provides strength and elasticity, while elastin enables tissue flexibility. These proteins are also present in bones, tendons, and ligaments, contributing to their mechanical stability and function.

Collagen fibers form a dense network that ensures the structural integrity of tissues, while elastin fibers enhance resilience and adaptability. Together, these proteins maintain the balance between strength and flexibility, supporting the body's physical demands.

Functional Proteins

Functional proteins include enzymes, which accelerate biochemical reactions, and transport proteins such as hemoglobin, which carries oxygen in the blood. Enzymes are highly specific and catalyze all vital chemical reactions in the body. Transport proteins like hemoglobin are essential for oxygen delivery and maintaining the acid-base balance in the bloodstream.

Examples of enzymes include amylase and lipase, which aid in the digestion of food, and DNA polymerase, which facilitates

DNA replication and repair. These proteins are indispensable for sustaining life and promoting cellular health.

Regulatory Proteins

Regulatory proteins, such as hormones, play pivotal roles in metabolic processes. Insulin, for example, regulates blood glucose levels by facilitating glucose uptake into cells for energy production.

Produced by the pancreas, insulin ensures glucose homeostasis and supports energy metabolism. Other regulatory proteins include growth hormones and neurotransmitters, which enable communication between nerve cells. These proteins orchestrate a range of physiological processes, including growth, development, reproduction, and metabolism, ensuring harmonious body function.

1.1.3. Carbohydrates

Carbohydrates are the body's preferred energy source, providing approximately 4 kcal per gram. They exist in two main forms: simple sugars and complex polysaccharides. Carbohydrates are essential for the rapid provision of energy, playing a critical role during physical activity and daily metabolism.

Simple Sugars

Glucose, fructose, and galactose are readily available energy sources. Glucose is particularly vital for powering the brain and muscles. It is absorbed from food and transported through the bloodstream.

The regulation of blood glucose levels is crucial for maintaining homeostasis and is controlled by hormones like insulin and glucagon. Glucose can be immediately used for energy production within cells or stored as glycogen in the liver and muscles for quick mobilization when energy demands increase.

Complex Carbohydrates

Starch and glycogen serve as storage molecules. Glycogen is stored in the liver and muscles, acting as a rapid energy reserve during physical activity. It can be converted back into glucose when the body requires additional energy.

The storage and mobilization of glycogen are regulated by enzymes such as glycogen synthase and glycogen phosphorylase, which respond to hormonal and neural signals. Starch, found in plant-based foods, is broken down by digestive enzymes into simple sugars, which are then absorbed into the bloodstream.

Conclusion

The chemical composition of the human body reflects its remarkable adaptability and efficiency in utilizing macronutrients to sustain life. Fats, proteins, and carbohydrates work in harmony to meet energy demands, support structural integrity, and regulate critical physiological processes. Understanding these components provides insight into the intricate balance that underpins human health and survival, offering a foundation for exploring the broader themes of energy transformation and the life cycle in subsequent chapters.

Energy and Transience -The Journey After Death

1.2. Energy metabolism in the body

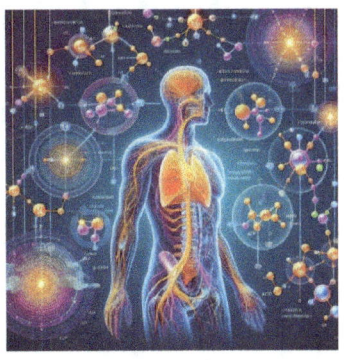

Energy metabolism encompasses the biochemical processes through which the body converts nutrients into energy and utilizes this energy to support its functions. This complex system is divided into two primary categories: **catabolism** and **anabolism.** These interconnected processes allow the body to adapt to changing energy demands and environmental conditions, ensuring optimal functioning and survival.

1.2.1. Catabolism

Catabolism refers to the breakdown of nutrients to generate energy. This process occurs in several key stages:

Glycolysis

Glycolysis is the initial step in glucose metabolism, occurring in the cytoplasm of cells. During this process, glucose is broken down into pyruvate, producing two molecules of ATP per glucose molecule. Additionally, NADH molecules are generated, which are later used in the electron transport chain. Glycolysis is tightly regulated by key enzymes, such as hexokinase and phosphofructokinase, which are controlled by feedback mechanisms. This regulation enables the body to precisely adjust glucose breakdown based on its current energy needs.

In addition to its role in energy production, glycolysis prepares glucose for the subsequent phases of metabolism, linking cellular energy production to the broader metabolic network. The efficiency of glycolysis is critical in meeting both immediate and long-term energy demands, particularly during activities requiring rapid energy, such as intense physical exertion.

Citric Acid Cycle (Krebs Cycle)

Within the mitochondria, pyruvate undergoes further breakdown in the citric acid cycle, a central hub of energy metabolism. This cycle produces high-energy molecules, including NADH and $FADH_2$, which feed into the electron transport chain for ATP synthesis. Beyond energy production, the citric acid cycle serves as an intersection point for the metabolism of fats and amino acids, highlighting its versatility.

The citric acid cycle is intricately regulated by substrate availability and the cell's energy needs. Enzymes catalyze each step, and the cycle's efficiency depends on the availability of coenzymes and substrates. This metabolic pathway illustrates the body's capacity to integrate multiple energy sources, ensuring a seamless energy supply regardless of dietary variations.

Oxidative Phosphorylation

The final stage of catabolism, oxidative phosphorylation, takes place in the inner mitochondrial membrane. Here, NADH and $FADH_2$ donate electrons to the electron transport chain. As electrons travel through the chain, energy is harnessed to synthesize ATP, the primary energy currency of cells. Water is produced as a byproduct, emphasizing the precision and balance of this process.

The efficiency of oxidative phosphorylation is governed by the proton gradient across the inner mitochondrial membrane, which drives the ATP synthase enzyme. This gradient represents stored potential energy, enabling the synthesis of ATP from ADP and inorganic phosphate. Oxidative phosphorylation is indispensable for maintaining cellular functions, supporting energy-demanding processes such as muscle contraction, neural activity, and tissue repair.

Expanding the Concept

Catabolism exemplifies the intricate coordination required for energy production within the human body. Each phase builds upon the previous, forming a tightly regulated system that responds dynamically to the body's energy demands. This process not only sustains vital functions but also ensures adaptability, allowing the body to thrive in diverse environmental and physical conditions.

Moreover, these processes underscore the efficiency of biological systems, where every molecule, enzyme, and cofactor plays a crucial role. Understanding catabolism provides insight into how the body utilizes nutrients, adapts to energy shortages, and maintains cellular integrity.

The next section will explore **anabolism**, the constructive counterpart to catabolism, completing the dynamic balance of energy metabolism and further illustrating the sophistication of the human energy system.

1.2.2. Anabolism

Anabolism is the process of synthesizing complex molecules from simpler building blocks, a process that requires energy. This constructive phase of metabolism is essential for growth, repair, and maintaining the structural and functional integrity of the body.

Protein Synthesis

Protein synthesis involves assembling amino acids into proteins, a highly energy-intensive process vital for cell repair, growth, and the production of enzymes. This process begins with the **transcription** of DNA into mRNA, which carries genetic instructions to the ribosomes. At the ribosomes, **translation** occurs, where the mRNA sequence is decoded into polypeptide chains.

These chains are then folded and modified into their final functional forms through intricate post-translational processes. A variety of enzymes and molecular machines ensure the precision and efficiency of this process, minimizing errors and maintaining cellular health. Protein synthesis underpins virtually all cellular activities, from muscle repair to the production of antibodies in the immune response.

Glycogenesis

Glycogenesis is the formation of glycogen from glucose molecules, a process that occurs in the liver and muscles to store excess glucose for later use. This storage mechanism allows the body to manage energy availability effectively, ensuring that glucose can be rapidly mobilized during periods of increased energy demand.

The enzyme **glycogen synthase** catalyzes glycogen formation and is activated by insulin, which signals the presence of excess glucose. When energy is needed, glycogen is broken down into glucose through glycogenolysis, ensuring a quick energy supply. This tightly regulated system highlights the body's ability to balance energy storage and utilization, adapting to varying physiological demands.

Lipogenesis

Lipogenesis converts excess glucose and other nutrients into fatty acids, which are stored as triglycerides for long-term energy reserves. This process primarily occurs in the liver and is regulated by enzymes such as **acetyl-CoA carboxylase** and **fatty acid synthase.**

Hormonal signals play a key role in lipogenesis. **Insulin** promotes this process during periods of energy surplus, while hormones like glucagon and adrenaline inhibit it during energy shortages. Lipogenesis ensures that surplus energy is efficiently stored and can be mobilized during times of fasting or increased energy demands. This long-term storage capability is particularly critical for survival during periods of food scarcity or high physical exertion.

Conclusion

The human body functions as a complex and highly regulated energy system. Its ability to extract, store, and efficiently utilize energy from various nutrients is fundamental to survival and optimal functioning. Fats, proteins, and carbohydrates each play specific and indispensable roles in this intricate system, enabling the body to adapt to diverse conditions and maintain homeostasis.

These macronutrients support a wide range of physiological processes, ensuring health and vitality. Through precise regulation and seamless interplay, the body can meet both immediate and long-term energy demands. This dynamic balance underscores the sophistication of human metabolism and its ability to sustain life in a constantly changing environment.

1.3. Energy Balance and Physiological Functions

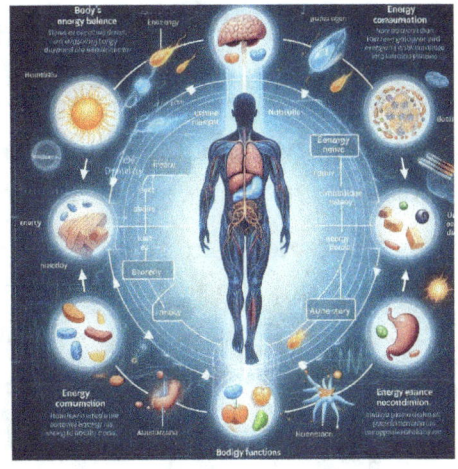

The energy balance of the human body is a complex equilibrium that reflects the delicate difference between energy intake and energy expenditure. To maintain essential bodily functions, the average adult requires a daily energy intake of approximately 2,500 to 3,000 kilocalories (kcal). This energy is derived from the consumption of food and beverages and must be carefully matched with energy expenditure to preserve physical health and overall well-being.

An imbalance between energy intake and expenditure can result in weight gain or loss, influencing the risk of various health conditions. The body's energy balance is composed of several interconnected components, each playing specific roles and governed by a range of factors. These components work harmoniously to ensure that the body performs all necessary functions efficiently. Understanding these components and their interactions is critical for grasping the body's energetic dynamics and the fundamental principles that sustain life and its functions.

1.3.1. Basal Metabolic Rate (BMR)

The basal metabolic rate (BMR) represents the amount of energy the body requires at complete rest to sustain basic life-supporting functions such as breathing, heartbeat, cellular metabolism, and maintaining body temperature. This energy is required 24/7, regardless of whether the body is active or resting. BMR accounts for approximately 60 to 75 percent of daily energy expenditure, underscoring its significant role in energy balance.

Several factors influence BMR, including age, sex, body composition, and genetic predisposition. These variables collectively determine how efficiently the body utilizes and expends energy at rest.

Age
As individuals age, their bodies tend to consume less energy at rest. This is primarily due to a reduction in muscle mass, which has a higher energy demand, and an increase in fat tissue, which requires less energy. Additionally, metabolic processes slow down over time, resulting in a lower BMR. This natural decline explains why older adults generally require fewer calories than younger individuals to maintain their weight.

Sex
Men typically have a higher BMR than women, largely due to their greater muscle mass. Muscle tissue consumes more energy than fat tissue, leading to higher caloric needs for men to sustain basic bodily functions. In contrast, women tend to have a higher proportion of fat tissue, which contributes to a lower BMR.

Body Composition
Individuals with a higher proportion of muscle mass exhibit a higher BMR, as muscle tissue is metabolically active and demands more energy than fat tissue. Athletes or those who engage in regular strength training often have elevated metabolic rates. A well-trained body not only burns more energy during physical activity but also at rest, as the maintenance of increased muscle mass requires additional energy.

Genetic Predisposition

Genetic makeup can significantly influence metabolic efficiency and BMR levels. Some individuals have a naturally faster metabolism due to their genetic predisposition, resulting in higher energy expenditure at rest. Conversely, others may have a slower metabolism, leading to a lower BMR. These genetic differences help explain why some people find it easier or harder to maintain their weight despite similar dietary and exercise habits.

Conclusion

The BMR is a cornerstone of the body's energy expenditure, reflecting how the body sustains vital functions even at rest. Factors such as age, sex, body composition, and genetics contribute to individual variations in metabolic rates. By understanding these influences, we can gain insights into the intricate mechanisms that govern energy balance and the body's ability to adapt to changing needs.

This foundational knowledge sets the stage for exploring other components of energy expenditure and their roles in maintaining health and vitality. The body's capacity to regulate its energy requirements dynamically is a testament to its remarkable complexity and efficiency.

1.3.2. Physical Activity

The energy expended through physical activity varies greatly depending on the type, intensity, and duration of the activity. Physical activity is a crucial component of energy balance and can significantly influence daily caloric expenditure. Regular exercise not only helps maintain physical fitness but also plays a vital role in weight management and disease prevention.

Light Activities

Light activities include daily movements such as walking, household chores, or light yoga. These activities require a moderate amount of energy and contribute to mobility and overall well-being. While they have a smaller impact on increasing daily caloric expenditure, they are essential for promoting an active lifestyle and mitigating the negative effects of a sedentary routine.

Light activities help boost metabolism and keep the body in motion, which is critical for maintaining health over time. Incorporating these activities into daily routines can reduce the risk of chronic illnesses associated with inactivity, such as cardiovascular diseases and metabolic disorders, and contribute to sustained physical and mental health.

Intense Activities

Intense physical activities, such as running, swimming, cycling, or weightlifting, demand significantly more energy and substantially increase daily caloric expenditure. These activities are effective in improving cardiovascular health, promoting muscle growth, and enhancing overall physical fitness.

By markedly raising the body's energy requirements, intense activities help burn excess calories and are particularly beneficial for weight control. Additionally, they improve endurance, strength, and flexibility, reducing the risk of chronic conditions such as heart disease, diabetes, and obesity. Engaging in regular, high-intensity exercise also supports mental health, fostering stress reduction and better sleep quality.

1.3.3. Thermogenesis

Thermogenesis refers to the body's heat production, which plays an essential role in energy balance. This process includes various mechanisms by which the body releases energy in the form of heat to maintain a constant internal temperature and support metabolic functions. Thermogenesis also contributes to burning excess energy, making it particularly relevant for weight regulation.

Digestive Thermogenesis

Also known as the **thermic effect of food (TEF)**, digestive thermogenesis describes the energy required for digesting, absorbing, transporting, and metabolizing food. After eating, energy expenditure increases as the body works to extract nutrients and convert them into usable energy or storage forms such as glycogen and fat.

The thermic effect varies depending on the type of food consumed. Proteins, for instance, have a higher thermic effect than fats and carbohydrates, meaning the body expends more energy to process proteins. TEF can account for up to 10% of daily energy expenditure and plays a significant role in regulating energy balance.

Active Thermogenesis

This form of thermogenesis involves heat production through physical activity and muscle work. Any form of movement, whether it be high-intensity exercise or everyday activities, raises body temperature and increases energy expenditure.

Active thermogenesis is a critical mechanism for burning excess energy and regulating body temperature during physical exertion. Combining active thermogenesis with regular exercise maximizes its positive effects on metabolism, leading to better energy balance control and improved overall health. This synergy underscores the importance of maintaining an active lifestyle for long-term physical and mental well-being.

Summary

The human body functions as a highly sophisticated energy system that utilizes various chemical compounds for energy production and storage. Key sources include fats, proteins, and carbohydrates, which are managed through comprehensive metabolic processes. The energy balance of the body comprises several key factors, such as basal metabolic rate (BMR), physical activity, and thermogenesis.

These components work in harmony to ensure the body performs all necessary functions efficiently. Understanding these processes is critical to comprehending the body's energetic dynamics and the principles that sustain life. A well-balanced energy system is vital for maintaining health, preventing over- or underweight conditions, and fostering long-term well-being.

Chapter 2:

The Final Moments of Life

The final moments of life are characterized by a multitude of profound and complex physiological and biochemical changes within the human body. These changes are integral to the natural process of dying, encompassing both physical and biochemical aspects. During this final phase, significant shifts occur in the functioning of organs, blood circulation, and respiration. Concurrently, profound biochemical adjustments prepare the body for its impending end.

For the individual experiencing these changes and their loved ones, this period is uniquely challenging. Emotional distress and physical transformations demand patience and understanding. The person may face a variety of symptoms that are both physically and emotionally taxing, while loved ones must navigate the difficulty of witnessing this process and providing necessary support, all while managing their own emotions.

A comprehensive understanding of the underlying physiological and biochemical processes can significantly ease this challenging time. Such knowledge offers deeper insights into the dying process itself, fostering empathy for the individual and their family. It enables caregivers and loved ones to provide appropriate support, alleviate fears, and approach the remaining time with greater composure and clarity. Understanding these changes and the mechanisms behind them allows caregivers and family members to better meet the needs of the individual, ensuring that this final chapter is approached with dignity and care.

2.1. Physiological Changes Immediately Before Death

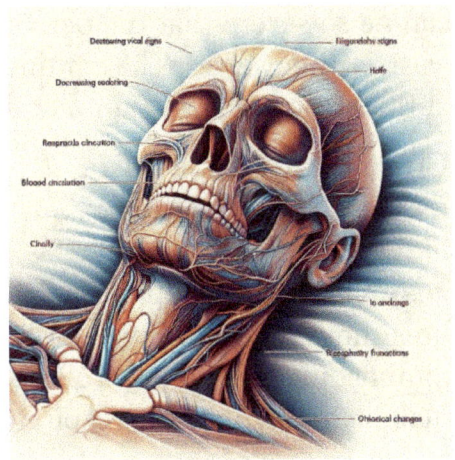

In the hours and days preceding death, the body undergoes a series of physiological changes that signal the end of life. While these changes vary from person to person, several well-documented signs and symptoms are commonly observed.

2.1.1. The Process of Dying

Dying is an exceedingly complex process that can be divided into several stages, each marked by distinct symptoms and signs that indicate the body is preparing for its conclusion.

Terminal Phase:

In the final hours to days before death, bodily functions slow significantly. Blood pressure steadily decreases, and the heartbeat may become irregular and weaker. Breathing often becomes shallower and more erratic. These changes are clear indicators that the body is gradually ceasing its vital functions.

Other symptoms during this phase may include a marked reduction in urine output and increasing fatigue or weakness. The individual may struggle to stay awake or conscious as vital body functions decline. The terminal phase is often accompanied by general weakness, making it difficult for the individual to perform simple daily activities. This weakness can be so severe that the person is unable to get out of bed or care for themselves. The continuous decline in bodily functions at this stage is a clear sign that death is imminent.

Changes in Breathing

Breathing patterns may become irregular, and some individuals exhibit what is known as **Cheyne-Stokes respiration.** This pattern involves periods of rapid, deep breathing followed by pauses. These patterns reflect the diminishing functionality of the central nervous system and its inability to regulate the respiratory muscles effectively.

In some cases, **agonal breathing** may occur, characterized by labored, irregular breaths, often accompanied by rattling sounds caused by secretions in the airways. These changes in breathing are often distressing for loved ones, as they signal the nearness of death. For the individual, these breathing patterns can cause discomfort, sometimes necessitating medications to ease respiration and improve comfort. Cheyne-Stokes and agonal breathing are unmistakable signs of the central nervous system's failure to maintain the body's vital functions, indicating that death is imminent.

Changes in the Cardiovascular System

The heartbeat slows and weakens, and in the final hours before death, the heart may exhibit intermittent pauses. Blood pressure can drop drastically, signaling that the cardiovascular system is nearing its final cessation.

This phase is often accompanied by a cooling of the extremities as the body redirects blood flow to vital organs, reducing circulation to peripheral tissues. This can lead to cold, bluish discoloration of the hands and feet, a phenomenon known as **cyanosis.** These changes in the cardiovascular system are clear indications that the body is relinquishing its efforts to sustain life and is transitioning toward death.

For many individuals, a noticeable decrease in heart rate

further underscores the cardiovascular system's inability to maintain effective blood flow. These changes can be distressing for the individual and their caregivers, making it crucial for healthcare providers and loved ones to offer understanding and support to ease this phase as much as possible.

Changes in Skin

The skin may appear pale and mottled due to reduced blood circulation. In the final hours, a bluish tint may develop on the hands, feet, and lips, a sign of oxygen deprivation in the blood and declining circulatory efficiency.

The skin often becomes cooler and may feel moist or clammy, further indicating the diminished functionality of the circulatory system and the impending physical changes associated with death. These skin changes can be particularly alarming for loved ones, as they are visible indicators of the body's transition toward death.

It is vital for caregivers and family members to provide compassionate and empathetic care during this time, helping the individual and their loved ones navigate these visible and emotional signs with as much comfort and understanding as possible.

Summary

The final moments of life involve a range of profound physiological changes that prepare the body for its eventual cessation. From alterations in breathing and circulation to shifts in skin color and temperature, these changes are clear indicators of the body's transition.

Understanding these processes can help caregivers, healthcare professionals, and family members provide appropriate care and emotional support. By recognizing these signs, it becomes possible to approach this final chapter with greater sensitivity, ensuring that the individual experiences dignity and comfort in their last moments.

2.1.2. Neurological Changes

Neurological changes are a significant aspect of the final moments of life. As the central nervous system begins to fail, a range of symptoms emerge that affect the body's overall function. These changes have both physical and emotional impacts and often serve as clear indicators of imminent death.

Changes in Consciousness

The individual may become increasingly confused, drowsy, or slip into a coma. These alterations in consciousness indicate that the brain is no longer functioning efficiently, and the ability to regulate bodily functions is diminishing. The person may stop speaking and become unresponsive to verbal cues or physical contact.

Hallucinations or visions of deceased loved ones are not uncommon and can be viewed as part of the transitional process. These experiences are often intense and emotional for both the individual and their loved ones. Confusion and disorientation can be frightening for the individual, making it essential for caregivers and family members to provide a calming and supportive presence to promote comfort and reassurance.

Responsiveness to Stimuli

Reflexes, such as the pupil's reaction to light, may become diminished or disappear entirely. Responsiveness to external stimuli decreases, and the eyes may appear fixed or glassy. This is a sign that neurological functions are deteriorating and the brain is losing its ability to control the body.

The gag reflex may also weaken, increasing the risk of aspiration if liquids or food enter the airways. The reduced responsiveness highlights the central nervous system's declining capacity to maintain basic functions. These changes are often unmistakable and can be particularly distressing for loved ones, as they starkly illustrate the individual's deteriorating condition.

Physical Movements

Involuntary movements or muscle twitches may occur, but coordination and control over voluntary movements diminish. These muscle twitches, known as **myoclonus**, often signal the declining influence of the central nervous system on the body.

In some cases, seizures may occur due to uncontrolled neuronal activity in the brain. These neurological changes indicate that the body is preparing for death, losing control over essential functions. For both the individual and their loved ones, these involuntary movements can be unsettling, underscoring the progressive nature of the dying process.

Summary

Neurological changes in the final moments of life reflect the central nervous system's diminishing ability to regulate and control the body. From shifts in consciousness and reduced reflexes to involuntary movements, these changes are clear indicators of the body's transition.

Understanding these signs can help caregivers and family members navigate this difficult time with greater empathy and preparedness. By recognizing and responding to these changes with sensitivity, they can provide the individual with comfort and dignity in their final hours.

Energy and Transience -The Journey After Death

2.2. Biochemical Processes at the Time of Death

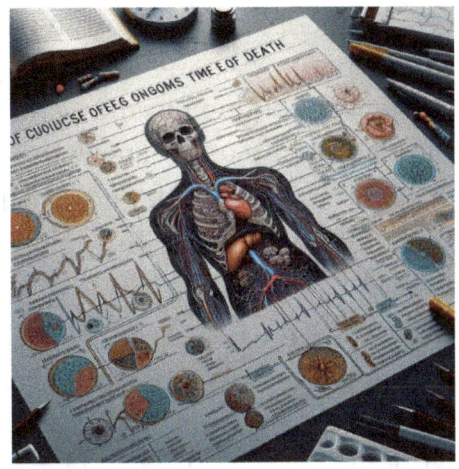

At the moment of death, various biochemical processes begin to unfold, driving the profound changes within the body. These processes mark the transition from a living system to a state of decomposition and are crucial for understanding death as a biological phenomenon.

2.2.1. Energy Depletion and Metabolic Changes

Energy Failure

When the heart ceases to beat and blood circulation halts, the supply of oxygen to cells is interrupted. Without oxygen, cells can no longer produce the essential molecule adenosine triphosphate (ATP), which serves as the primary energy source for cellular functions. ATP is indispensable for all biochemical processes within cells, and its absence causes cellular systems to fail rapidly.

The loss of energy leads to a breakdown in the integrity of cell membranes, critically impairing cell structure and function. This triggers a cascade of biochemical changes that ultimately result in cell death.

The process of energy failure is intricate, involving multiple stages. The mitochondria, known as the powerhouses of cells, cease functioning due to the lack of oxygen required for ATP production. As a result, vital processes such as ion pump activity in cell membranes can no longer be maintained. This disruption in ion balance destabilizes the cell membranes, rendering the cells unable to perform their functions. This marks the initial step in a chain of biochemical events leading to complete cell death.

Metabolic Changes

Following the cessation of oxygen supply, the body shifts to anaerobic metabolism to break down the remaining glucose without oxygen. This process, called anaerobic glycolysis, produces lactic acid as a byproduct. The accumulation of lactic acid results in increasing acidity in tissues, a condition known as lactic acidosis. This acidity disrupts the function of many enzymes and proteins within cells, further compromising cellular activity.

The rising concentration of lactic acid contributes significantly to systemic acidosis. This acidosis exacerbates cellular damage and accelerates the decomposition of tissues. The biochemical changes initiated by energy depletion and metabolic shifts lead to the structural disintegration of cells, hastening the body's overall decomposition.

This complex sequence of biochemical events provides critical insight into how death occurs at the cellular level and how subsequent changes affect the body as a whole.

Summary

In summary, death triggers a cascade of biochemical events, starting with an immediate depletion of energy. The lack of oxygen halts ATP production, causing cellular functions to collapse. Simultaneously, anaerobic metabolism produces lactic acid, leading to acidosis, which further damages cells and accelerates tissue breakdown.

These interconnected biochemical processes set the stage for the body's decomposition. Understanding these mechanisms is vital for comprehending the intricate dynamics of death and the subsequent changes that affect the body scientifically.

2.2.2. Onset of Autolysis

Autolysis

Once cells exhaust their energy reserves and lose the ability to produce new energy, a remarkable and inevitable process begins: autolysis. In this state of absolute energy deprivation, cells start to break themselves down. Enzymes, typically safely contained within the lysosomes of cells, are now released. These enzymes, which normally play critical roles in digesting cellular waste and maintaining cellular health, begin to decompose the cell from within.

Autolysis is the initial phase of cellular decomposition and the subsequent breakdown of tissues. During autolysis, intracellular enzymes are activated and systematically dismantle cellular structures. Cell membranes become permeable, allowing these enzymes to digest cellular components, such as proteins, lipids, and nucleic acids. This self-digestion results in the loss of cellular structure and the gradual collapse of tissue integrity.

The process of autolysis varies in speed across different tissues and organs, depending on factors such as temperature, pH levels, and the enzymatic makeup of the cells. Organs with high concentrations of enzymes, such as the liver and pancreas, undergo autolysis particularly rapidly. The release and activation of lysosomal enzymes occur almost immediately after death, triggering a cascade of reactions that eventually lead to the complete breakdown of cellular structures.

As autolysis progresses, the physical and chemical properties of tissues change. Affected cells and tissues become softer and lose structural integrity. This process is a critical aspect of natural decomposition, preparing the body for subsequent stages of microbial decomposition. Autolysis represents the initial phase of postmortem decay and is essential for understanding the biological processes that occur after death.

Summary

Autolysis marks a crucial transition in the postmortem process. Through the release and activation of lysosomal enzymes, cells begin to digest themselves, leading to the destruction of tissue structure. This process initiates a sequence of events that ultimately transitions the body into a state of full biological decomposition. Understanding autolysis and its mechanisms is fundamental to the scientific study of death and the subsequent decomposition processes.

2.2.3. Role of Enzymes

Enzymatic Activity

Enzymes play a central and indispensable role in the decomposition of cellular structures after death. Key enzymes involved include **proteases** and **lipases**, each performing specific functions.

Proteases:

These enzymes break down proteins by cleaving peptide bonds between amino acids. This process degrades structural proteins and enzymes within cells, significantly compromising cellular integrity and function.

Lipases:

These enzymes are responsible for breaking down fats. They catalyze the breakdown of triglycerides into fatty acids and glycerol, releasing lipid components into surrounding tissues.

The activity of these enzymes has profound effects on the decomposition process. By degrading proteins and fats, cell membranes become increasingly permeable, rendering cells incapable of maintaining their internal environment. This increased permeability eventually leads to cell rupture, a phenomenon known as **cell lysis.** The released cellular components, including organelles and other intracellular materials, spill into surrounding tissues, further accelerating decomposition.

Enzymatic activity is not confined to individual cells; it also affects entire tissues and organs. The release and action of proteases and lipases disrupt the microstructure of tissues, contributing to the gradual disintegration of the body's anatomical integrity. Organs with dense cellular structures and high enzymatic activity exhibit accelerated decomposition, underscoring the significance of enzymes in the breakdown of the body postmortem.

Summary

The final moments of life are characterized by a series of complex physiological and biochemical changes that mark the transition to death. These changes affect both physical and neurological functions, forming an integral part of the natural dying process.

Energy collapse following cardiac arrest, subsequent metabolic shifts, and the onset of autolysis are pivotal steps in this transition. The release of lysosomal enzymes, coupled with the activity of proteases and lipases, leads to the breakdown of cellular structures and tissue disintegration. These biochemical processes amplify one another, accelerating the body's decomposition.

Understanding these intricate processes provides valuable insights into the final phase of life and the biological mechanisms governing the transition from life to death. This knowledge can help caregivers and loved ones navigate this challenging time with greater empathy and preparedness, offering the best possible support to the individual.

Ultimately, these insights contribute to viewing death not merely as an end but as a part of a natural and unavoidable biological cycle. This understanding can help demystify the dying process, fostering a calmer and more dignified approach to life's inevitable conclusion.

Chapter 3:

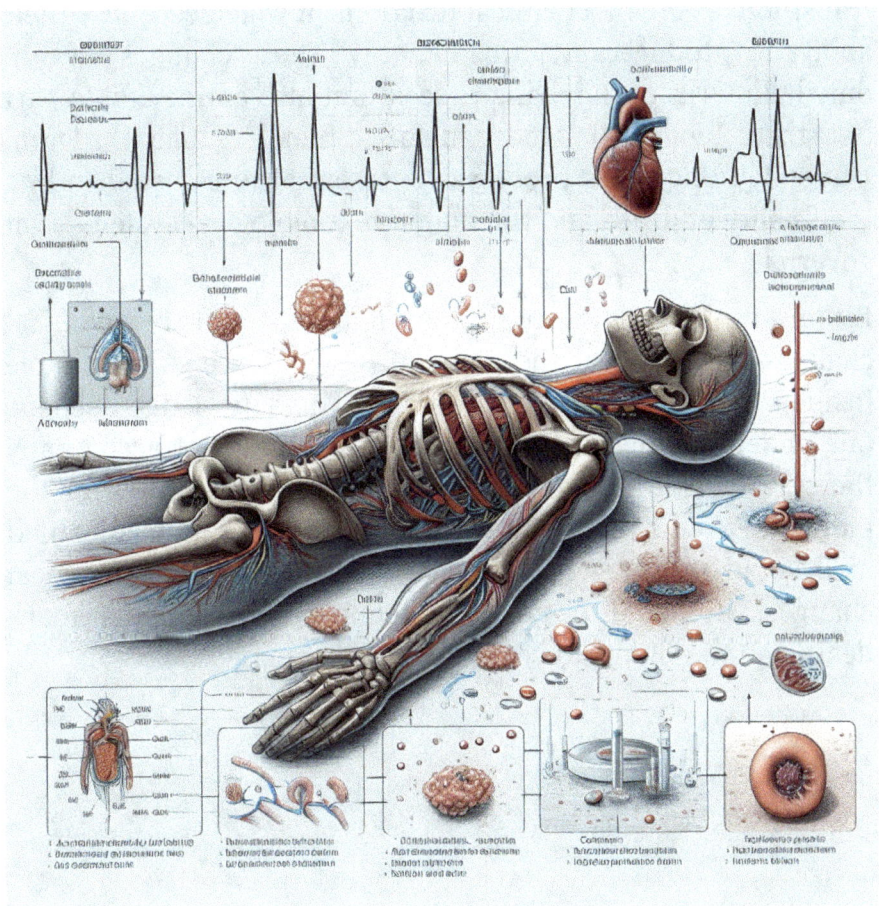

At the Moment of Death

The moment of death is a profoundly significant and transformative event, marking the end of life and characterized by a complex interplay of physiological and biochemical changes. This final moment represents the ultimate cessation of the biological processes that sustained life until this point. During this transition, definitive changes occur at both the organ and molecular levels.

Physiological processes, such as the cessation of cardiac activity and alterations in respiration, are integral to this transition, as are biochemical reactions in which cells lose their ability to produce and utilize energy. These changes are not only hallmarks of the dying process but also an inevitable part of the biological mechanisms that regulate death. Understanding these processes is essential to comprehending the finality of death and the underlying biological systems that govern it.

By examining the specific physiological and biochemical changes that occur at the moment of death, we gain valuable insights into the nature of life's end. This knowledge deepens our understanding of the intricate processes that accompany the cessation of life and fosters a greater appreciation for the biological events that define this transition. Through a detailed analysis of the final biochemical reactions and physiological changes, we can better grasp the sequence of events that mark death and the significance of this critical moment.

3.1. Defining Death

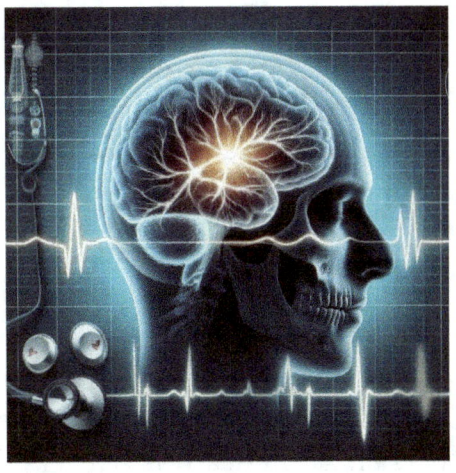

Death is generally defined as the irreversible cessation of all signs of life. This concept encompasses two primary aspects: **clinical death** and **biological death**.

Clinical death occurs when the heartbeat and respiration have ceased. At this stage, the heart is no longer maintaining circulation, and the lungs have stopped exchanging oxygen and carbon dioxide. Often considered the first recognizable moment of death, clinical death may still be reversible in some cases through immediate medical intervention, such as resuscitation. However, without successful intervention, this temporary state inevitably progresses to biological death.

Biological death represents the final phase, during which all cellular functions have completely stopped. At this stage, cells sustain irreversible damage and can no longer carry out vital processes. This includes the failure of ATP production, loss of cell membrane integrity, and the onset of autolysis. Biological death signifies the body's total loss of functionality as a living system, with no possibility of restoring life.

Medical and Legal Perspectives

The definition of death varies slightly across medical and legal contexts, yet there is a broad consensus that death marks the irreversible cessation of life-sustaining functions. In modern medicine, **brain death** is frequently used as the definitive criterion for death. Brain death refers to the complete and permanent cessation of all brain activity, including that of the brainstem.

This condition is confirmed through various tests that demonstrate the absence of brain activity, such as electroencephalography (EEG), which records no electrical signals. Brain death is distinct from other conditions, such as vegetative states or comas, where some brain functions may persist. Brain death is irreversible and is legally recognized as death, even if machines artificially maintain heartbeat and respiration.

This definition carries significant ethical and legal implications, particularly concerning organ transplantation and the cessation of medical treatments. It underscores the need for precise diagnostic criteria to determine the exact point of death.

Ethical and Practical Implications

Understanding these definitions and criteria is vital for medical practice and the ethical decisions associated with them. It allows physicians to accurately determine the time of death and take appropriate actions, whether ending life-sustaining measures or preparing for organ donation. Additionally, these criteria provide clear guidelines for family members and caregivers to recognize the loss and begin the process of saying goodbye.

Summary

The moment of death is characterized by a series of complex physiological and biochemical changes. The transition from clinical death to biological death marks the final loss of life functions, defined with precision in modern medical contexts such as brain death.

Understanding these processes and definitions is essential for comprehending the finality of death and addressing the associated medical, ethical, and legal challenges. By exploring these transitions, we gain a deeper appreciation of the intricate biological mechanisms that delineate life from death.

3.1.1. Clinical Death

Cardio-Respiratory Death

Cardio-respiratory death occurs when the heart stops beating and breathing ceases. It is the most immediate and evident indication of death. At this stage, resuscitation may still be successful if prompt medical intervention is provided. The primary method of intervention is **cardiopulmonary resuscitation (CPR)**, which involves manual chest compressions and artificial ventilation to maintain blood circulation and oxygen delivery.

The critical time window for successful resuscitation is typically just a few minutes after cardiac and respiratory arrest. The faster resuscitation measures are initiated, the higher the likelihood of successfully restoring life functions. With every passing minute without intervention, the chances of survival diminish significantly.

Diagnosis

Clinical death is diagnosed by the absence of heartbeat, breathing, and pulse. Medical personnel use a range of techniques and devices to confirm these signs. Commonly employed tools include stethoscopes to listen for cardiac and respiratory activity and heart monitors to detect electrical activity in the heart.

An **electrocardiogram (ECG)** is a vital tool in this process, recording the heart's electrical signals to determine whether it is still beating. A flatline on the ECG indicates the absence of any cardiac activity.

In addition to monitoring the heart and lungs, medical staff assess for the absence of reflexes that are typically present in living individuals. The **pupillary light reflex** is a key test, where the pupils fail to constrict in response to light if clinical death has occurred. Other reflexes, such as the **corneal reflex** (the eyelid's response to corneal stimulation) and pain reflexes, are also evaluated.

These diagnostic measures are essential to accurately and reliably confirm clinical death. They allow healthcare providers to make informed decisions regarding the next steps, whether continuing resuscitation efforts or transitioning to palliative care. In many cases, clinical death is confirmed by multiple medical professionals to ensure diagnostic accuracy and avoid prematurely halting life-saving measures when there is still a chance of survival.

Clinical vs. Biological Death

The distinction between clinical and biological death is crucial in medical practice. While clinical death may still be reversible under certain conditions, biological death marks the point at which all cellular and systemic functions have irreversibly ceased. Understanding this difference is vital not only for executing appropriate medical actions but also for effectively communicating with family members and making decisions in ethically and legally sensitive scenarios.

Summary

Clinical death is defined by the absence of heartbeat, respiration, and pulse, confirmed through various diagnostic methods. Quick and precise resuscitation measures are essential to maximize the chances of successfully restoring life functions. Understanding and accurately diagnosing clinical death are fundamental aspects of medical practice and critical for appropriate management in emergency situations.

3.1.2. Biological Death
Irreversible Cellular Death
Once clinical death has occurred and resuscitation efforts prove unsuccessful, the process of irreversible cellular death begins. Biological death is reached when all cellular activities, including ATP production and protein synthesis, cease entirely. At this stage, the cells can no longer sustain metabolic functions, produce energy, or synthesize essential proteins. This state signifies that the body's tissues are no longer viable, and cellular life cannot be sustained. Biological death is thus the final cessation of all physiological and biochemical processes that support life.

Mechanisms of Cellular Death
Biological death is driven by several cellular processes, primarily **autolysis** and **necrotic decomposition**:

Autolysis
Autolysis, or self-digestion, involves the breakdown of cells by their own enzymes. Lysosomes, cellular organelles housing digestive enzymes, play a central role in this process. Normally, lysosomes degrade damaged or unnecessary cellular components in living cells. After death, however, the lysosomal membranes become permeable, releasing enzymes such as proteases and lipases into the cytoplasm. These enzymes dismantle proteins, lipids, and other macromolecules, progressively dissolving cellular structures. Autolysis is a key natural process in postmortem tissue decomposition and marks the early stages of biological disintegration.

Necrosis
Necrosis is a key mechanism of cell death during biological death, characterized by uncontrolled cellular destruction caused by external factors such as toxins, infections, or physical trauma. Unlike apoptosis, a programmed and

orderly process of cell death, necrosis is abrupt and disorganized. In this process, cells swell as their internal balance is disrupted, and their membranes become compromised. This loss of membrane integrity leads to the sudden collapse of cellular structures.

One hallmark of necrosis is the release of intracellular components into the surrounding tissues. When the cellular membrane ruptures, these contents, including enzymes and organelle debris, spill into the extracellular environment. This uncontrolled release triggers inflammatory responses in the surrounding tissue, often exacerbating tissue damage and accelerating the breakdown process.

Necrosis differs from autolysis, where cells self-destruct via internal mechanisms. While autolysis is an intrinsic process driven by lysosomal enzymes, necrosis is extrinsically induced and is often accompanied by collateral damage to adjacent cells and tissues. For instance, in postmortem conditions, necrosis can occur due to environmental factors such as bacterial invasion or physical disruptions to tissue integrity.

Together with autolysis, necrosis plays a central role in dismantling the body's cellular framework during biological death. These processes contribute to the irreversible disintegration of tissues, marking a critical phase in the transition from life to decomposition.

From Clinical to Biological Death

The transition from clinical death to biological death is a continuous process involving successive stages of cellular degradation. While clinical death may still be reversible with timely medical intervention, biological death signifies the complete and irreversible cessation of life-supporting functions. Understanding these mechanisms is essential for various fields, including transplantation medicine, forensic pathology, and scientific studies of postmortem processes.

Summary

Biological death is characterized by the irreversible cessation of all cellular activities. Processes like autolysis and necrosis play central roles in dismantling cellular structures and breaking down tissues. This knowledge not only aids in understanding the progression of death but also provides critical insights into scientific and medical practices surrounding the end of life.

Energy and Transience -The Journey After Death

3.2. Immediate Pphysiological Reactions at the Moment of Death

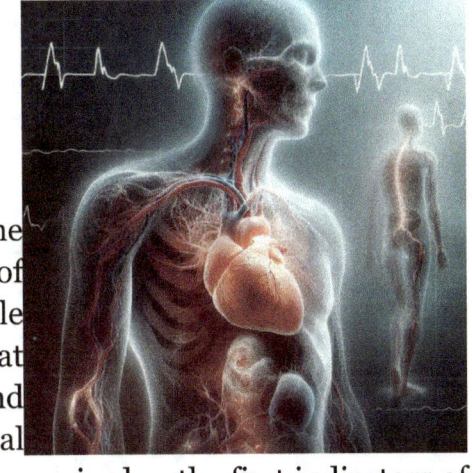

The moment death occurs, the body undergoes a series of intricate and inevitable physiological changes that signal the cessation of life and the end of all biological functions. These processes, recognized as the first indicators of death, occur systematically at both cellular and systemic levels. Among the most significant are the cessation of cardiovascular activity, the halt of respiratory functions, and the initial breakdown of bodily tissues. Together, these changes initiate an irreversible cascade that progresses over the hours and days following death.

A key consequence of death is the abrupt loss of oxygen supply and blood circulation. Without the delivery of oxygen, cells are unable to sustain their critical metabolic processes. This results in the rapid cessation of cellular activity, particularly in organs like the brain, which is highly dependent on a continuous oxygen supply. The brain stops functioning within minutes, causing an immediate and complete loss of consciousness and signaling the definitive end of vital processes.

At the same time, the lack of oxygen and energy triggers internal degradation. Enzymes normally tasked with digesting food within the gastrointestinal system now begin to break down the body's own tissues in a process known as autolysis. This marks the start of decomposition, as cells and tissues lose their structural integrity and begin to dissolve. These changes are not only consequences of death but also active contributors to the body's progressive breakdown.

This intricate interplay of physiological reactions underscores the finality of death. It signifies the transition from a living organism to a system in decay, initiating a natural sequence that ultimately returns the body's materials to the environment. These processes, while complex, are a fundamental part of the life cycle, illustrating the interconnectedness of life and death.

3.2.1. Cardiovascular System
Cardiac Arrest

The cessation of the heart's activity leads to an immediate halt in blood circulation. This event terminates the transport of oxygen and nutrients to tissues via the arteries and veins. As a result, cellular metabolism ceases rapidly due to the abrupt lack of oxygen and energy, causing an accumulation of metabolic byproducts such as carbon dioxide and lactic acid. This buildup results in tissue acidosis, which accelerates cell death and initiates the decomposition process.

The cardiovascular system, with the heart at its core, is responsible for maintaining the continuous delivery of oxygen and nutrients throughout the body. When the heart stops beating, this vital function comes to an immediate halt. Tissues that once relied on a steady supply of oxygen and nutrients face a sudden and critical deprivation. This oxygen deprivation, or hypoxia, damages cells irreversibly, as they lose their ability to generate energy via aerobic respiration.

In the moments following cardiac arrest, cells shift to anaerobic metabolism in an attempt to meet energy demands without oxygen. However, this process produces lactic acid, which lowers the pH in tissues and results in acidosis. This acidic environment disrupts cellular functions and accelerates the breakdown of cells. Simultaneously, carbon dioxide, which can no longer be expelled through respiration, accumulates in the blood, further exacerbating acidosis and compounding cellular degradation.

Blood Clotting

With the cessation of blood flow, coagulation begins within the vessels. Blood clotting results in the formation of clots that settle within the circulatory system. Concurrently, gravity causes blood to pool in the lowest parts of the body, leading to the formation of livor mortis, or postmortem discoloration. These purplish-red discolorations are caused by blood accumulating in capillaries and small veins of dependent body areas. Livor mortis is a clear indication that blood circulation has ceased and is often used to estimate the time of death.

Blood clotting initiates minutes after the heart stops. In the absence of circulation, blood settles due to gravity, creating the visible patches of livor mortis. These patches begin forming within 20 to 30 minutes postmortem and become clearly discernible after 2 to 4 hours. Forensic pathologists utilize these discolorations to deduce both the time of death and whether the body has been moved postmortem. Changes in the location or appearance of livor mortis provide critical clues during criminal investigations.

The clotting of blood within the vessels can also influence subsequent decomposition processes. Clots may obstruct the flow of residual fluids and impact bacterial activity during tissue breakdown. This interplay between coagulation and decomposition underlines the interconnected nature of postmortem changes.

Summary

The cessation of the cardiovascular system marks the first phase of the body's immediate reactions to death. Cardiac arrest halts oxygen delivery, disrupts metabolism, and initiates tissue acidosis, while blood pooling and clotting result in visible changes like livor mortis. These processes are critical for forensic professionals, providing insights into the time and circumstances of death. Understanding these intricate changes aids accurate death diagnosis and lays the foundation for examining subsequent decomposition stages.

3.2.2. Respiration

Respiratory Arrest:
The cessation of breathing immediately following death results in a rapid and significant drop in blood oxygen levels, a condition known as hypoxemia. This oxygen deficit has profound effects on the body's tissues and cells, triggering a cascade of physiological and biochemical disruptions. Respiration typically facilitates the exchange of oxygen and carbon dioxide: oxygen enters the bloodstream to sustain cellular metabolism, while carbon dioxide, a metabolic waste product, is expelled. When this critical exchange halts, carbon dioxide levels in the blood rise dramatically, disrupting the body's delicate acid-base balance.

Deprived of oxygen, tissues and cells can no longer sustain aerobic metabolism, forcing a shift to anaerobic pathways to meet their energy demands. This results in the production of lactic acid, which lowers the pH of tissues, leading to acidosis. The accumulation of acid impairs cellular functions, destabilizing enzymes and biochemical processes essential for maintaining cellular integrity. Organs like the brain, highly sensitive to oxygen deprivation, begin to fail within minutes, initiating the breakdown of biological functions and accelerating the onset of decomposition.

Carbondioxide accumulation:
As breathing stops, carbon dioxide—ordinarily expelled through respiration—builds up in the blood. This rapid accumulation triggers respiratory acidosis, a condition where elevated levels of carbon dioxide cause blood pH to drop. Carbon dioxide, present in the blood as carbonic acid,

exacerbates the acidity, disrupting enzymatic activities and cellular mechanisms that rely on stable pH levels.

The rising acidity from carbon dioxide buildup further accelerates tissue damage and cell death. Combined with hypoxemia, this creates an environment where cellular structures begin to break down, releasing their contents into surrounding tissues. This process intensifies the systemic progression toward decomposition, as cells lose their integrity and biochemical pathways collapse under the compounding stress.

Summary:
Respiratory arrest is one of the most immediate and recognizable physiological signs of death, initiating a sequence of critical changes. Oxygen deprivation and carbon dioxide accumulation disrupt cellular homeostasis, triggering acidosis and tissue damage that hasten decomposition. These early postmortem changes are key to understanding the physiological aftermath of death and provide important forensic clues about the timing and circumstances surrounding death. The cessation of respiration marks the start of an irreversible transformation that bridges the transition from life to death.

3.2.3. Physical Changes
Muscle Relaxation and Rigor Mortis
Immediately after death, the body undergoes a series of physical changes that affect both the muscles and the overall body structure. At the initial stage of this process, the muscles relax completely, resulting in a general limpness of the body. This immediate effect of death occurs as the electrical activity that typically stimulates muscle contraction ceases. Without continuous nerve stimulation and the electrical impulses necessary for contraction, the muscles enter a state of complete relaxation.

However, this initial muscle relaxation is soon followed by the process of rigor mortis. Rigor mortis, commonly referred to as postmortem rigidity, is the stiffening of the muscles, typically beginning 2 to 6 hours after death. This process arises because the muscles, deprived of ATP (adenosine triphosphate), can no longer relax. ATP serves as the critical energy source required for muscle relaxation. When the heart stops beating and blood circulation ceases, ATP production halts, leading to an accumulation of active myosin heads, which bind to actin filaments and lock the muscle fibers into a contracted state. The absence of energy prevents the muscles from relaxing, causing the body to stiffen progressively.

Rigor mortis typically begins in the smaller muscle groups, such as those around the eyes and jaw, before spreading to the larger muscle groups. This rigidity peaks approximately 12 to 24 hours after death and may persist for up to 72 hours, depending on factors such as ambient temperature and the individual's body composition. Following the peak of rigor mortis, the muscles gradually relax again as cellular breakdown continues. Enzymes that facilitate the decomposition of muscle proteins become active, leading to a secondary state of relaxation as decomposition progresses.

Body Temperature Decline (Algor Mortis)

Another critical physical process that occurs after death is the decline in body temperature, also known as algor mortis. Following the cessation of vital functions, the body begins to adjust to the ambient temperature. This cooling process occurs naturally as the body can no longer regulate its internal heat and gradually matches the surrounding environmental temperature.

The rate at which the body cools can vary significantly, influenced by numerous factors. These include the ambient temperature, the clothing worn by the deceased, the individual's size and body composition, and the humidity levels in the environment. In cooler surroundings, the body cools more rapidly, while in warmer or more humid conditions, the cooling process is slower. As a general guideline, the body temperature tends to drop by approximately 1 to 1.5 degrees Celsius per hour during the initial hours after death until it reaches equilibrium with the ambient temperature.

The decline in body temperature provides valuable clues for estimating the time of death, especially in forensic investigations. By measuring the body temperature and comparing it with the surrounding temperature, it is possible to estimate the time of death with reasonable accuracy. The progression of algor mortis also serves as an indicator of other postmortem changes, assisting forensic experts in determining not only the time but also the potential circumstances of death.

Summary

In summary, the physical changes that occur after death are intricate processes involving muscle relaxation, the onset and resolution of rigor mortis, and the decline in body temperature (algor mortis). These physiological phenomena mark the transition from life to death and offer critical insights for medical and forensic analysis. Understanding these physical changes provides valuable information about the time of death and the underlying causes, contributing significantly to the fields of forensic science and postmortem investigation.

Energy and Transience - The Journey After Death

3.3. Chemical and biochemical changes

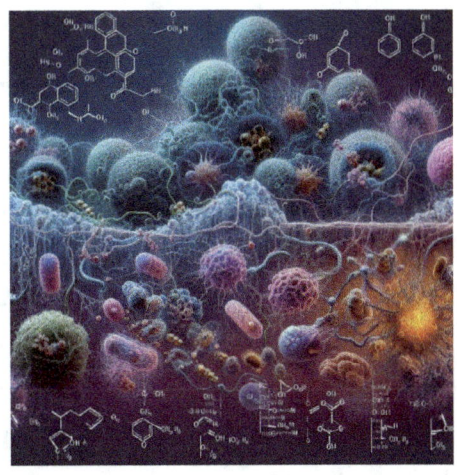

Once death occurs, the body undergoes a profound cascade of chemical and biochemical changes that pave the way for the subsequent decomposition process. These processes are fundamental to understanding postmortem changes and illustrate the body's transition into a state of decomposition. They are crucial for analyzing the biological and chemical mechanisms that accompany death and the subsequent decay process.

3.3.1. Loss of Energy

ATP Depletion

With the cessation of the heartbeat and the end of respiration, the production of ATP (adenosine triphosphate) within the body immediately halts. ATP serves as the primary energy source essential for a multitude of cellular processes, including ion transport and the stabilization of cell membranes. Without the continuous production of ATP, an immediate energy deficit occurs within the cells, leading to severe consequences for cellular integrity and function.

Cell membranes, typically maintained by active transport mechanisms fueled by ATP, destabilize under these conditions. The depletion of ATP results in the loss of membrane structure and functionality. This disruption triggers a cascade of cellular damage that accelerates cell death.

The lack of ATP also halts critical processes such as ion transport across cellular membranes and the maintenance of membrane potential. This disruption in ion balance, particularly concerning sodium and potassium, exacerbates cellular membrane damage and hastens cell death. As cells lose the ability to sustain their normal operations, they succumb to a chain reaction of damage, culminating in the complete breakdown of cellular structures.

Cell Membrane Degeneration

In the absence of the energy provided by ATP, cell membranes begin to disintegrate. This breakdown permits the release of intracellular enzymes and other cellular contents into the surrounding tissue, initiating a process known as autolysis. Autolysis, or self-digestion, marks the onset of cellular breakdown and accelerates tissue decomposition.

The lysosomal enzymes, typically contained within the cells, are released during this process and begin degrading the cell from within. This enzymatic activity dissolves proteins, lipids, and other cellular components, gradually dismantling the structural integrity of the cells. As these cellular contents and enzymes diffuse into the surrounding tissue, they catalyze further decomposition and expedite the overall decay process.

Summary

The chemical and biochemical changes that commence immediately after death are pivotal to the decomposition process. ATP depletion and subsequent cell membrane degeneration are central to the transition from a living state to a state of decay. These processes offer critical insights into postmortem decomposition and the biological mechanisms underlying death and its accompanying transformations within the body.

3.3.2. Acid-Base Balance
Acidosis
After death, dramatic chemical changes occur within the body, primarily driven by a lack of oxygen and an accumulation of carbon dioxide. These changes result in significant acidification of the blood and tissues. The drop in blood pH, known as acidosis, has profound effects on the biochemical and enzymatic processes within cells. Elevated acidity accelerates the enzymatic breakdown of cellular components. Enzymes, typically responsible for cellular metabolism and repair processes, function optimally within a specific pH range. A lower pH severely impairs their function, hastening the degradation of cellular structures and functions.

The resulting acidic environment exacerbates cellular deterioration as many proteins and enzymes lose their structure and functionality. This intensifies the natural decomposition process, accelerating the breakdown of cells and tissues. Acidosis also affects the integrity of cellular membranes, impairing the cells' ability to maintain their normal functions. This process facilitates rapid decomposition and the progression of biological changes following death.

Electrolyte Imbalances
Alongside acidosis, disturbances in the body's electrolyte balance occur. Electrolytes such as sodium, potassium, calcium, and magnesium are crucial for maintaining cellular functions, particularly in muscle and nerve tissues. After death, as cell membranes lose their integrity and ATP levels drop to zero, electrolytes begin to redistribute uncontrollably. The loss of electrolytes, such as sodium and potassium, further disrupts cellular functions. These imbalances affect the electrical activity within cells, exacerbating postmortem changes.

The disruption of electrolyte balance compromises the ability of cells to transmit electrical signals, a critical function in muscle and nerve tissues. This dysfunction contributes to further tissue degradation, as the cellular electrical activity necessary for normal functions, such as contractions and nerve impulses, ceases. The resulting imbalance exacerbates the already progressing decomposition, accelerating the physical changes that follow death.

Summary

In summary, changes in acid-base balance and electrolyte disturbances illustrate the profound biochemical processes that unfold immediately after death. The drop in pH caused by acidosis and the disruptions in electrolyte balance are critical for understanding postmortem decomposition. These changes offer valuable insights into the biological mechanisms driving the breakdown of the body after death. Together, they are key factors that accelerate the decay process and the deterioration of cellular and tissue functionality in the postmortem state.

3.3.3. Enzymatic Decomposition

Autolytic Enzymes

Following death, a profound process of enzymatic decomposition begins within the body, laying the foundation for the subsequent breakdown of tissues. During life, many enzymes responsible for cellular metabolism and other essential processes are encapsulated within specialized organelles called lysosomes. These enzymes are adept at breaking down various biological molecules to maintain normal cellular function. However, with the onset of death and the subsequent loss of cellular integrity, these enzymes are released from their lysosomal compartments.

As cellular membranes lose their structural integrity, lysosomal enzymes freely invade the intracellular environment. Key enzymes such as proteases, lipases, and nucleases systematically dismantle cellular structures.

Proteases degrade proteins into smaller peptides and amino acids, lipases break down lipids (fats), and nucleases target nucleic acids such as DNA and RNA. This enzymatic activity initiates the widespread disassembly of cellular components, including membranes, organelles, and other structural elements.

This process, known as autolysis, represents a form of self-digestion in which cells are broken down by their own enzymes. With energy deficits rendering cellular membranes unstable, regulatory barriers that previously controlled enzymatic activity are lost, triggering an uncontrolled degradation process. The enzymatic breakdown of cellular structures destroys cellular integrity, contributing significantly to the rapid and widespread decomposition of tissues. Autolysis is a critical phase of postmortem decomposition, offering key insights into how the body begins to disintegrate after death.

Onset of Microbial Activity

In addition to autolytic enzymes, microbes naturally present in the body play a central role in advancing decomposition after death. Shortly after death, bacteria residing in the digestive tract, skin, and other bodily regions begin to invade tissues. These microbes include various species of bacteria, fungi, and other microorganisms that typically coexist symbiotically within the body.

With the cessation of bodily defenses against microbial growth, these organisms thrive in the favorable conditions created by death. They exploit decaying tissues as a nutrient source, accelerating the breakdown of cellular structures. During this process, microbes release gases such as methane, hydrogen sulfide, and ammonia, which contribute to the characteristic odors of decomposition.

Microbial activity induces further changes in tissues. The accumulation of gases within the body causes visible bloating, while microbial digestion continues to degrade tissues, accelerating the overall decomposition process and intensifying the distinct signs of putrefaction. This microbial activity is a crucial factor in altering tissue structure after death, driving the comprehensive disintegration of the body.

Summary

The process of enzymatic decomposition after death is intricate and multifaceted. Autolysis, driven by the body's own cellular enzymes, and microbial decomposition, facilitated by bacteria and fungi, are central mechanisms of postmortem tissue breakdown. Together, these processes orchestrate the progressive disintegration of tissues and are essential for understanding the biological changes that follow death. Knowledge of these enzymatic and microbial processes offers valuable insights into the natural mechanisms governing the body's transition after life.

3.4. The transition to decomposition

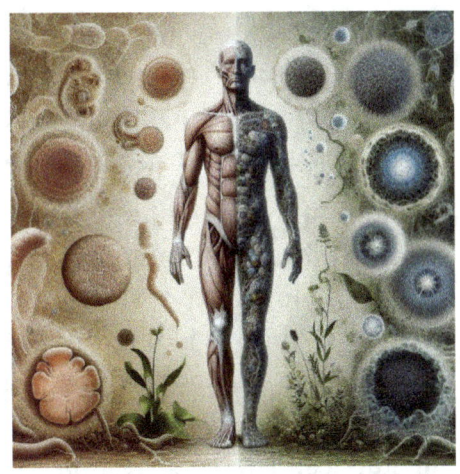

The transition from a state of life to one of decomposition is a profound and multifaceted process that begins immediately after death. This transition marks the cessation of all biological and physiological functions and leads to a complex series of processes that result in the breakdown of the body. It represents a phase characterized by a variety of visible and invisible changes that gradually transform the appearance and composition of the body. These changes occur not only at a surface level but also penetrate deeply into the biochemical and mechanistic aspects of bodily functions.

The decomposition process unfolds in distinct stages, each marked by specific mechanical and biochemical events that collectively define the entire progression. These phases are interconnected, with each stage influencing and building upon the previous one. Gaining a comprehensive understanding of these phases is essential for appreciating the profound transformations that occur as life transitions into decomposition.

This process begins with the immediate cessation of all vital biological processes and functions that were active prior to death. Breathing stops, the heart ceases to beat, and blood circulation halts, initiating an irreversible sequence of decomposition. This transformation is not merely physical but also involves significant chemical and biological shifts, driving the body through several stages of change.

In the hours immediately following death, the early stages of decomposition commence, as the body transitions from the vitality of life into a state of decay. These initial stages are marked by both visible and tangible changes, resulting from the absence of essential physiological processes and the accumulation of metabolic waste products and residual substances. These early phases are crucial for understanding decomposition, as they lay the groundwork for subsequent, deeper biochemical and physiological transformations.

A thorough understanding of this transitional process is key to comprehending the complete progression from life to decomposition. This journey manifests in a wide range of changes occurring at both the cellular and structural levels. The complexity and breadth of these processes underscore the profound biological transformations that take place after death.

This chapter explores these stages in detail, offering insights into the mechanisms that govern the transition from life to decay. By delving into the early, intermediate, and advanced phases of decomposition, we aim to illuminate the intricate processes that define the end of life and the return of the body's components to the natural cycle.

3.4.1. Early Decomposition: Rigor Mortis

Rigor mortis, commonly referred to as postmortem rigidity, is a striking and unmistakable sign that begins shortly after death and signifies a profound transformation within the body. This phase of early decomposition typically sets in within 2 to 6 hours after death and marks a critical stage in the postmortem changes of the body. Rigor mortis is the first visible and physical manifestation indicating the transition of the body from a living state to one of decomposition.

In this initial phase, a progressive stiffening of the muscles begins, spreading throughout the body and reaching its peak approximately 12 to 24 hours after death. This stiffening is caused by a biochemical imbalance within the muscle cells, triggered by the absence of the vital energy source ATP (adenosine triphosphate). During life, ATP is essential for muscle function, allowing muscle fibers to relax after contraction by disengaging the actin and myosin filaments responsible for muscle movement.

The mechanism behind this stiffening is complex and involves several biochemical steps. In a living body, ATP-dependent processes ensure that muscle fibers return to their relaxed state after contraction. This continuous supply of ATP is crucial for maintaining muscle flexibility and functionality. Following death, ATP production ceases abruptly, causing the existing actin-myosin complexes to remain locked in a contracted state. As muscles are no longer able to relax, rigidity gradually develops and spreads throughout the body.

The stiffness of rigor mortis typically persists for about 24 to 48 hours. During this period, the muscles become fully rigid, resulting in a noticeable change in the physical appearance of the body. This rigidity serves as a clear indicator of death and the onset of decomposition. As decomposition progresses, muscle proteins begin to break down due to advancing biochemical and enzymatic processes. This degradation eventually softens the muscles, causing the initial stiffness to dissipate. This phase is significant in postmortem changes, as it can provide critical insights into the progression of decomposition.

The rate and pattern of rigor mortis development can be influenced by various environmental factors. Ambient temperature plays a crucial role; higher temperatures accelerate biochemical reactions, leading to a faster onset of rigor mortis. In warmer environments, stiffness tends to develop more quickly, whereas cooler temperatures can slow the process. Humidity levels in the surroundings also affect the progression of rigor mortis, with moist conditions potentially speeding up muscle stiffening. Additionally, the physical activity of the deceased prior to death can impact the rate and intensity of rigor mortis. Individuals who were physically active may exhibit a different rate of muscle stiffening due to distinct biochemical conditions in their muscles compared to less active individuals.

These complex interactions highlight the multifaceted nature of postmortem changes and the numerous factors influencing the course of early decomposition. Understanding rigor mortis provides critical insights into the timeline of postmortem events and helps forensic and medical professionals determine the time and conditions surrounding death.

3.4.2. Postmortem Lividity (Livor Mortis)

Immediately after death, a complex process begins in the body, marked by a series of physiological and biochemical changes. One of the earliest and most visually recognizable changes is the appearance of postmortem lividity, also known as livor mortis. These discolorations result from the pooling of blood in the lower regions of the body due to the cessation of circulation. Once the heart stops beating and blood circulation halts, gravity causes blood to collect in the lowest parts of the body. This process typically begins within 20 to 120 minutes after death and is a clear indication of the onset of decomposition.

Mechanism of Blood Pooling

The formation of livor mortis is directly linked to the sudden cessation of the circulatory system. Under normal circumstances, blood is continuously pumped throughout the body, delivering oxygen and nutrients to tissues and removing waste products. After death, this critical pumping action ceases. Blood vessels retain their structural integrity for a short time, allowing gravity to pull the blood downward. In these lower regions, the blood begins to pool, unable to circulate any further. This gravitational effect intensifies the blood accumulation, leading to characteristic purplish discolorations on the skin in areas where the body is closest to the ground.

Initially, the color of lividity may appear as a red or bluish tint, which darkens over time due to further decomposition of the blood. This discoloration typically develops within the first few hours postmortem and may deepen to a dark blue or nearly black shade, depending on environmental conditions and the progression of decomposition. The specific pattern and timing of this color change can be influenced by various factors, including ambient temperature, humidity, and the body's position after death.

Influence of Environmental Conditions

The rate and distribution of livor mortis can be significantly affected by environmental factors. Temperature plays a key role in determining how quickly lividity forms and how pronounced it becomes. In warmer environments, the pooling of blood occurs more rapidly and with greater intensity due to the acceleration of biochemical reactions within the body. Conversely, cooler temperatures slow this process, as the body's internal reactions and blood flow decelerate. Humidity also influences livor mortis by affecting the rate of fluid evaporation from the body, which can impact the development of these discolorations.

The posture of the deceased at the time of death further affects the pattern of livor mortis. The way the body is positioned determines where blood accumulates under the influence of gravity. For instance, if a person dies lying in a certain position, blood will pool more prominently in the areas of the body in closest contact with the ground. This information is invaluable for forensic investigations, as it helps reconstruct the body's position at the time of death.

Forensic Significance

Livor mortis holds substantial forensic value, offering critical insights into the circumstances of death. Forensic experts analyze the pattern and coloration of lividity to infer the body's position postmortem and potentially identify whether the body was moved after death. The characteristic discolorations provide clues to the deceased's orientation and can help investigators determine if the original position of the body was altered.

Moreover, livor mortis is a useful indicator of the time of death. By closely examining the changes in pattern and color over time, forensic pathologists can estimate the postmortem interval. These observations enable experts to approximate how long the body has been deceased, contributing significantly to the timeline of events surrounding the death.

In summary, livor mortis is an indispensable tool in forensic pathology. It not only aids in determining the precise cause and time of death but also reveals crucial details about the circumstances surrounding the death. Understanding the formation and forensic relevance of these discolorations is essential for accurately reconstructing the events leading to death and for resolving forensic investigations.

3.4.3. Microbial Decomposition

The process of microbial decomposition begins immediately after death and plays a crucial role in transitioning the body from a living organism to a state of decay. This transformation is driven by the intense activity of microbes that naturally reside in the body and specialize in breaking down organic substances. These microbes, predominantly bacteria and fungi, are found in various areas of the body, including the gastrointestinal tract, skin, and mucosal surfaces. Their activity initiates a series of biochemical and physical changes that are essential to the progressive decomposition of the body.

Gas Formation and Bloating

Mechanism of Gas Formation

After death, microbes within the gastrointestinal tract and on body surfaces begin to focus on breaking down the body's organic materials. Many of these microbes are anaerobic bacteria, which thrive in environments devoid of oxygen. They metabolize the remaining nutrients in the body, producing a variety of gases as metabolic byproducts. The primary gases generated during microbial decomposition include methane, hydrogen, hydrogen sulfide, and carbon dioxide.

These gases gradually accumulate in body cavities and tissues, as the usual pathways for gas exchange and transport cease to function after death. Gas production often begins within the first 24 hours postmortem and continues to increase as microbial activity persists. The buildup of gas pressure leads to visible bloating, particularly in the abdominal region and other low-lying areas of the body. This bloating often becomes pronounced, with gases visibly distending the skin, and serves as a clear marker of advancing decomposition.

Impact on Body Structure

The gas-induced bloating significantly affects the body's structural integrity. As the gas pressure builds, the skin and underlying tissues are stretched to their limits. This extreme distension can cause visible tears and ruptures in the skin, as it is unable to withstand the elevated internal pressure. These ruptures are not merely superficial; they indicate profound changes within the internal tissues, which are also compromised by the gas pressure and decomposition.

In advanced stages of microbial decomposition, the pressure can become so intense that the skin bursts. These ruptures allow for the release of gases and cellular materials into the surrounding environment, further accelerating the decay process. The release of gases and other decomposition products contributes to environmental contamination as these substances disperse into the area around the body.

The accumulation of gases and resultant bloating also compromise the integrity of internal organs. The pressure exerted on internal structures leads to additional damage and tissue changes, promoting further disintegration of the body. The bloating exacerbates organ damage, hastening their breakdown and contributing to the overall collapse of bodily structures.

Summary

In summary, microbial decomposition, particularly the formation of gases and associated bloating, constitutes a central component of the decomposition process. The production and accumulation of gases within the body not only result in visible effects but also cause profound structural changes. Understanding these processes is critical for recognizing the progression of postmortem decomposition and gaining deeper insights into the complex biochemical and physical transformations occurring during this phase.

3.4.4. Odor Formation

As decomposition progresses, the body develops characteristic odors caused by the release of a variety of volatile organic compounds. These compounds, including cadaverine and putrescine, are the primary contributors to the distinctive "smell of death," often described as unpleasant and pervasive. The formation of these odors is a complex interplay of microbial activity and chemical reactions that begin shortly after death and intensify as decomposition advances.

Mechanism of Odor Formation

The characteristic odor of decomposition arises directly from the microbial breakdown of organic molecules, particularly amino acids and proteins. After death, microbes naturally present in the human body initiate the degradation of remaining organic materials. This decomposition is largely driven by anaerobic bacteria, which thrive in oxygen-deprived environments. During this phase, various volatile compounds are produced, many of which are noted for their strong and unpleasant odors.

Cadaverine and putrescine are two key compounds responsible for the distinctive smell of a decaying body. Both are amines produced through the microbial degradation of lysine and ornithine. Lysine is an essential amino acid commonly found in proteins, while ornithine is an intermediate in the urea cycle. As proteins break down, microbial enzymatic activity converts these amino acids into cadaverine and putrescine.

Cadaverine and putrescine emit potent, foul odors often associated with the decomposition process. In addition to these primary compounds, other volatile organic compounds, such as methanethiol, hydrogen sulfide, and dimethyl sulfide, contribute to the overall smell. Methanethiol has a sulfurous odor reminiscent of rotting onions, while hydrogen sulfide smells like rotten eggs. Dimethyl sulfide further amplifies the unpleasantness of the odor, enhancing the overall stench associated with decomposition.

Impact on the Environment

The odors generated during decomposition can have significant environmental effects, particularly in confined or poorly ventilated spaces. In such environments, the smell can become particularly intense and persistent, posing challenges for the storage and handling of remains. In enclosed areas such as unventilated rooms or sealed containers, the odor can quickly accumulate, creating a pervasive and overwhelming atmosphere.

The widespread diffusion of decomposition odors can also impact surrounding environments. For example, if a body is found in a populated area like an apartment or building, the odor can lead to considerable discomfort and health concerns for nearby residents. The smell may penetrate walls and doors, spreading into adjacent rooms or buildings, necessitating additional measures to control and mitigate the odor. Common strategies to manage these odors include the use of air purifiers, odor neutralizers, and enhanced ventilation systems.

In forensic contexts, the smell of decomposition provides valuable information. The intensity and characteristics of the odor can offer clues about the postmortem interval and the condition of the remains. Forensic experts often analyze these odor properties to refine estimates of the time since death and to reconstruct details about the circumstances of death. The odor may also play a role in investigations, helping determine whether the death occurred in a confined space or under conditions that influenced odor dispersal.

Summary

The process of odor formation during decomposition is a complex and multifaceted aspect of decay, characterized by the production of volatile organic compounds such as cadaverine and putrescine. These compounds result from the microbial breakdown of amino acids and other organic molecules. The resulting odor affects not only the immediate environment of the remains but can also have broader environmental implications, particularly in confined or poorly ventilated areas. A thorough understanding of odor formation and its effects is essential for managing decomposition and conducting forensic analyses of postmortem changes.

Chapter 4:

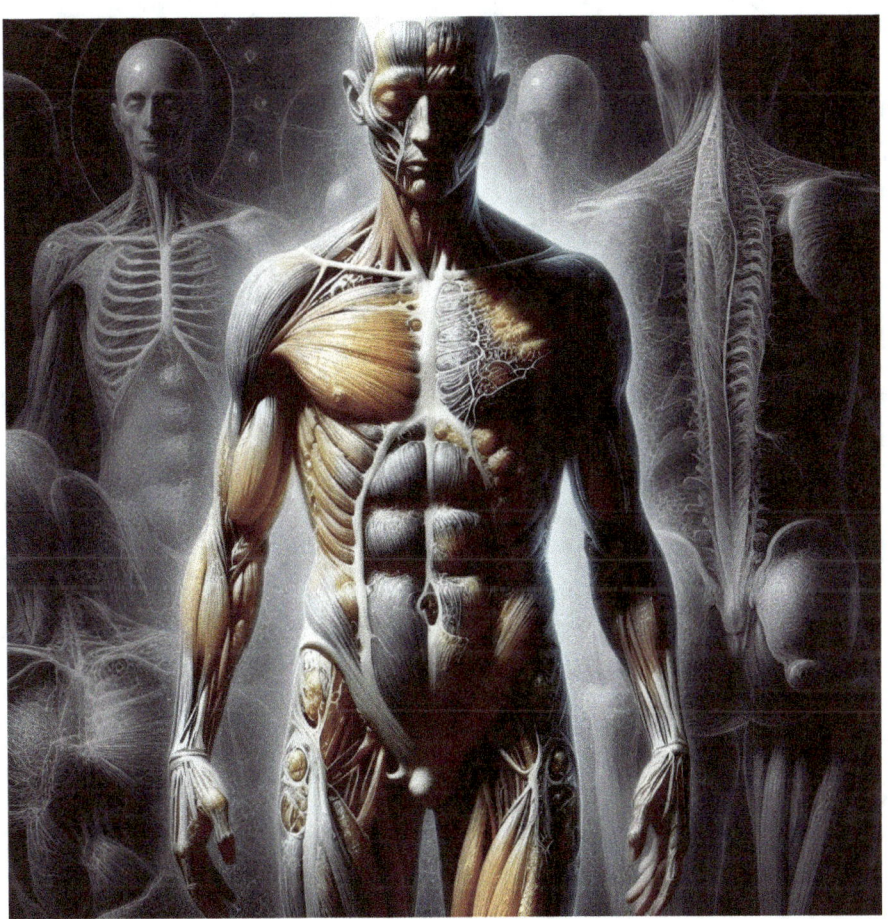

The first hours after death

4.1. Early physiological changes

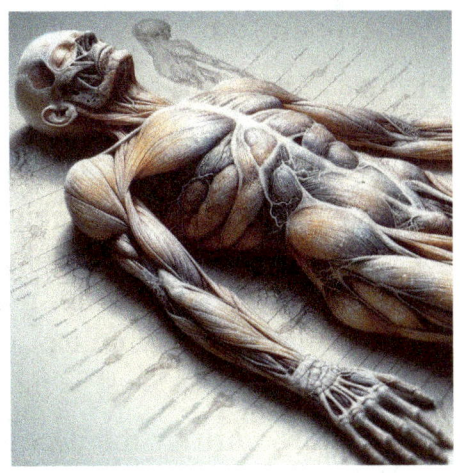

Immediately following death, the body undergoes profound physiological processes signaling the transition from life to decomposition. These changes, which range from visible phenomena to microscopic processes, set the stage for subsequent decomposition and provide crucial information for forensic investigations. The first hours postmortem are critical, as they establish the foundation for later stages of decay and offer insights into the mechanisms driving the body's progression toward complete decomposition. A thorough understanding of these early changes is essential for comprehending the entire trajectory from physiological alterations to the eventual breakdown of the body.

4.1.1. Temperature Changes

Algor Mortis

Upon death, the body ceases its heat production as metabolic processes responsible for generating warmth come to a halt. This cessation results in the body's gradual release of heat to the surrounding environment, a process known as *algor mortis*. The cooling of the body generally occurs at a rate of approximately 1 to 1.5 degrees Celsius per hour until it equilibrates with the ambient temperature. However, the pace of this cooling process is subject to various influencing factors.

Factors Influencing Temperature Changes

A range of environmental and physiological variables can impact the rate at which the body loses heat postmortem. The most significant factor is ambient temperature, which dictates the degree of heat loss. In cooler environments, the body cools more rapidly due to a greater temperature differential between the body and its surroundings. Conversely, in warmer settings, the cooling process slows, as the temperature gap diminishes.

The clothing worn by the deceased also plays a pivotal role in modulating heat loss. Thick clothing acts as insulation, slowing the cooling process, whereas thin or damp clothing can accelerate it. Storage conditions, such as the surface on which the body rests, further affect heat dissipation. For example, a body on a cold, hard surface loses heat more quickly than one placed on a softer, warmer surface.

Body mass significantly influences the rate of heat loss. Larger bodies with greater tissue and muscle mass retain heat longer due to their higher thermal capacity, while smaller bodies cool more rapidly. Additionally, the position of the body can impact cooling. Greater contact with a cold surface accelerates heat loss, whereas limited contact reduces the rate of cooling.

Practical Applications

In forensic medicine, analyzing the cooling of the body is invaluable for determining the time of death. By measuring the core body temperature and considering the factors outlined above, investigators and forensic pathologists can estimate the postmortem interval (PMI). This method is particularly useful in investigations or legal contexts where determining the precise time of death is critical.

However, estimating the time of death based on body temperature is not without its challenges. Environmental conditions, clothing, and other factors can significantly influence cooling rates, potentially leading to varying conclusions. Therefore, understanding *algor mortis* is only one component of a comprehensive forensic approach. To achieve the most accurate determination of the time of death, additional methods and techniques must be employed. A well-rounded grasp of these processes and their variations is essential for forensic professionals to draw precise and reliable conclusions.

Summary

The early physiological changes following death encompass a multitude of factors that influence subsequent decomposition processes. Understanding *algor mortis* and the associated temperature changes is vital for grasping the transition from life to decomposition and addressing the forensic challenges that arise when determining the time of death.

4.1.2. Blood Clotting

Onset of Clotting

Immediately following death, the body undergoes dramatic changes driven by various physiological processes. One of the first significant changes is blood clotting. With the cessation of the continuous blood flow maintained by the heart during life, the conditions within the body rapidly alter. Blood, which previously circulated through the vessels, begins to coagulate inside the vessels once circulation stops. This process, known as postmortem blood clotting, typically occurs within a few hours after death and is a direct result of the cessation of cardiac activity and blood flow.

Clotting is further accelerated by gravity, which causes blood to settle in the lower regions of the body. This phenomenon leads to the formation of "postmortem clots," which accumulate in veins and arteries. These clots vary in size and consistency depending on the time elapsed since death and the prevailing conditions. The formation of such clots is an early and definitive indicator of death and an integral aspect of postmortem physiological changes.

Signs of Death

Postmortem blood clotting results in the appearance of characteristic discolorations known as *livor mortis* or postmortem lividity. These marks are caused by the pooling of blood in the lower regions of the body due to gravity. Once circulation ceases, blood settles in these areas, creating visible discoloration. These marks, which are typically red to violet, appear most prominently in areas corresponding to the body's lower sections, such as the back, thighs, or buttocks.

The distribution and appearance of these discolorations provide valuable insights into the body's position at the time of death. Forensic experts analyze these patterns to infer the circumstances surrounding the death. For example, the precise location of livor mortis can help reconstruct the body's position and identify any postmortem movement or manipulation. These analyses are crucial for forming a comprehensive understanding of the circumstances of death and detecting potential tampering with the body.

Changes in Blood Clotting

The speed and nature of blood clotting after death can be influenced by a variety of factors. These include the deceased's medical history, particularly conditions or medications affecting blood coagulation. For instance, disorders like hemophilia, which impair the clotting process, or the use of anticoagulants that reduce the blood's clotting ability, can significantly alter postmortem clotting dynamics.

Environmental factors, such as temperature and humidity, also play a role in the progression of blood clotting. In warm, humid conditions, clotting may occur faster or slower compared to cooler, drier settings. These external conditions can affect the chemical and biological processes driving clot formation, thereby influencing the texture and extent of the clots.

Forensic Relevance

The examination of livor mortis and postmortem blood clots is a critical aspect of forensic pathology. Analyzing these changes can provide valuable information about the circumstances surrounding death and any potential postmortem manipulation. Unusual patterns or the absence of livor mortis might indicate a change in body position after death or interference with the corpse.

Detailed observation and documentation of livor mortis and blood clots are essential for constructing an accurate picture of the circumstances of death. Forensic experts must carefully assess the timing and formation of these features and consider factors that might have influenced the clotting process. This information significantly aids in determining the cause and context of death and forms an indispensable component of forensic investigations.

4.1.3. Muscle Relaxation and Rigor Mortis

Muscle Relaxation:

Immediately following death, the body undergoes significant physiological changes, one of the first being the complete relaxation of muscles. This phase, known as muscle relaxation, is characterized by a pronounced slackness throughout the body. In the initial hours postmortem, all muscles are soft and pliable as they lose their ability to contract. This condition results from the absence of ATP (adenosine triphosphate), an essential energy source for muscle contraction and relaxation during life. Without ATP, the muscles are unable to sustain active processes, leading to a state of total relaxation.

This period of muscle relaxation is temporary, lasting only for a limited time. The duration can vary depending on environmental factors such as temperature and humidity. In cooler environments, muscle relaxation tends to persist longer, whereas warmer conditions accelerate the process. During this stage, the body remains soft and flexible, marking an essential transition phase from life to the onset of decomposition.

Onset of Rigor Mortis

Approximately 2 to 6 hours after death, the next critical process, known as rigor mortis or postmortem rigidity, begins. During this phase, the muscles gradually stiffen, a phenomenon caused by biochemical imbalances in the body after death. While alive, muscle contraction and relaxation rely on a continuous supply of ATP. After death, ATP production ceases, and the existing energy reserves are depleted. Consequently, muscle fibers remain locked in a contracted state due to the absence of chemical processes necessary for relaxation.

Maximum muscle rigidity is typically reached within 12 to 24 hours postmortem. During this period, the body becomes significantly stiff, severely restricting movement. Rigor mortis generally persists for 24 to 48 hours before gradually dissipating as decomposition progresses and the muscle structures begin to break down. This phase of muscle stiffening is a critical time frame for estimating the time of death and provides valuable information for forensic investigations.

Phases of Rigor Mortis

The process of rigor mortis unfolds in several distinct phases. Initially, a slight stiffness develops in the muscles, which intensifies progressively until maximum rigidity is achieved. This initial stiffness usually begins in smaller muscles, such as those of the face and hands, before spreading to larger muscle groups. After reaching its peak, rigor mortis begins to subside as tissues decompose and the structural integrity of muscle fibers deteriorates. The duration and progression of these phases can be influenced by various environmental factors, including temperature, humidity, and the deceased's physical condition.

Variations in the rate and pattern of rigor mortis can provide insights into specific circumstances at the time of death. Factors such as ambient temperature or physical activity before death can significantly affect the progression of rigor mortis. Detailed analysis of the phases and intensity of muscle stiffening offers forensic experts crucial clues for determining the time of death and identifying any external influences that may have affected the process.

Forensic Significance

The examination of rigor mortis is of significant importance in forensic medicine, providing essential information about the time and circumstances of death. The distribution and duration of stiffness in the body can indicate the position of the deceased at the time of death. These findings are critical for reconstructing the events surrounding the death and identifying signs of potential tampering with the body.

Irregularities in rigor mortis, such as asymmetrical stiffening or delayed onset, may point to specific medical conditions or unusual circumstances. A meticulous examination and documentation of rigor mortis are therefore indispensable for forensic analysis, ensuring a comprehensive understanding of the cause and circumstances of death while considering all relevant factors in the investigation.

Energy and Transience -The Journey After Death

4.2. Biochemical processes and enzymatic activities

The hours immediately following death are marked by specific biochemical and enzymatic processes that signal the onset of decomposition and offer profound insights into the molecular changes occurring within the body. This phase is crucial for understanding the biological processes that underpin the progression toward complete decomposition. The biochemical reactions during this time are both automatic and triggered by the absence of vital nutrients and oxygen.

4.2.1. Enzymatic Decomposition

Autolysis

The process of autolysis begins immediately after death and serves as a pivotal step in the decomposition process. During autolysis, enzymes contained within lysosomes—a type of organelle specialized in the breakdown of cellular waste and redundant components—are released. Under normal circumstances, these enzymes are tightly regulated within their respective lysosomes to prevent cellular damage. However, after death, cellular functions cease, metabolic processes halt, and these enzymes are liberated, initiating the systematic breakdown of cellular structures.

Autolysis is an automatic, self-perpetuating mechanism.

Without the regulatory control of living cells and in the absence of nutrients and oxygen, the enzymes begin to degrade key cellular structures, including cell walls, nuclei, and organelles. This internal degradation process occurs rapidly as enzymes immediately target cellular components such as proteins, lipids, and nucleic acids. The result is a progressive disintegration of cellular structures, laying the groundwork for the body's subsequent decomposition.

Cell Membrane Damage

A critical aspect of early decomposition is the damage to cell membranes, driven by the depletion of ATP (adenosine triphosphate). ATP is an essential energy source that maintains cell membrane integrity and supports energy-dependent processes necessary for cellular function and stability. With the cessation of ATP production after death and the rapid depletion of remaining ATP reserves, cell membranes lose their stability and structural integrity.

Without ATP, lysosomal enzymes infiltrate the intracellular environment, accelerating the breakdown of proteins, lipids, and nucleic acids. Proteases degrade protein structures, lipases act on lipids, and nucleases break down nucleic acids like DNA and RNA. This continuous enzymatic degradation exacerbates cellular decomposition, leading to a significant loss of structure and function within the body.

The combined effects of autolysis and cell membrane damage initiate a cascade of further decomposition processes that occur at the molecular level. These biochemical changes ultimately manifest as visible signs of decomposition. Understanding these early biochemical and enzymatic activities is critical to comprehending the full dynamics of decomposition and its temporal progression.

The Process of Autolysis

Autolysis is a highly intricate and multilayered phase of

postmortem decomposition that unfolds in successive stages. Initially, cellular organelles—particularly lysosomes—are destabilized, leading to the release of their enzymes. Lysosomes, which are small, membrane-bound organelles containing enzymes specialized in breaking down proteins, lipids, and other molecules, play a central role in this process.

After death, the loss of cellular vitality triggers lysosomes to release their enzymes into the cytoplasm. These enzymes then act directly on cellular structures, degrading protective barriers like cell membranes. This destruction allows the enzymes to access and dismantle critical organelles such as the nucleus, mitochondria, and endoplasmic reticulum. These organelles, essential for cellular function, are systematically broken down during autolysis.

Enzymes involved in autolysis include proteases, lipases, and nucleases, each targeting specific components of the cell. Proteases digest protein structures, lipases act on lipids, and nucleases degrade nucleic acids. The enzymatic breakdown of these components accelerates the disintegration of cellular structures and contributes to the progressive dissolution of cellular components. Notably, the rate of autolysis varies across different tissues, influenced by the concentration and activity of enzymes present. Tissues like the liver and heart, which possess high enzymatic activity, decompose more rapidly than tissues such as skin, which break down more slowly.

Influence of External Factors

The speed and intensity of autolysis are not solely dependent on internal enzymatic activity but are also significantly influenced by external factors, including temperature, humidity, and the presence of microorganisms.

Temperature

Higher temperatures accelerate enzymatic activity, as molecules gain kinetic energy, leading to faster reaction rates.

Consequently, decomposition proceeds more rapidly in warm environments, as enzymes become more active and cellular structures degrade more quickly. Conversely, lower temperatures reduce molecular motion, slowing enzymatic reactions and significantly delaying decomposition.

Humidity

Moist conditions also play a pivotal role in promoting autolysis. Water acts as a medium for enzymatic and microbial activity, facilitating the breakdown of tissues. In humid environments, bacteria and fungi thrive, accelerating the decomposition process through their enzymatic contributions.

Understanding how temperature, humidity, and microbial presence influence autolysis is crucial for accurately interpreting the timeline and progression of decomposition. These factors are not only integral to grasping the postmortem changes within the body but also hold significant relevance for forensic analysis and determining the circumstances surrounding death.

Acid-Base Balance

Increase in Acidity

Immediately after death, the body's natural metabolic pathways cease abruptly and are replaced by anaerobic metabolic processes. This shift occurs due to the sudden cessation of blood circulation and the resulting lack of oxygen supply. In this state, cells are forced to break down glucose without oxygen, a process known as anaerobic metabolism. This metabolic activity produces lactic acid as a byproduct of incomplete glucose processing. The continuous formation of lactic acid during this anaerobic phase leads to a significant increase in tissue acidity.

Lactic acid has a corrosive effect on cellular structures, damaging cell membranes and attacking the internal components of cells. This heightened acidity results in further structural degradation, as typically stable cell membranes and other cellular components succumb to the persistent acidic assault. The progressive destruction of cellular structures accelerates the overall decomposition process, undermining cellular integrity and hastening tissue breakdown. The rise in acidity is thus a pivotal factor that influences the pace of decomposition and propels tissue degeneration.

pH Reduction

The accumulation of lactic acid and other acidic metabolites in the tissue causes a significant drop in pH levels. The pH level, which measures the concentration of hydrogen ions, indicates how acidic or basic a solution is. A lower pH reflects higher acidity and has several consequences for tissue decomposition. Acidic conditions enhance the activity of lysosomal enzymes, which play a central role in breaking down cellular structures. These enzymes, including proteases, lipases, and nucleases, are especially active in acidic environments and accelerate the degradation of proteins, lipids, and nucleic acids.

The lowered pH therefore drives increased enzymatic activity, which further speeds up tissue decomposition. This intensification has broad implications for the rate and progression of decomposition, as enzymatic activity directly amplifies tissue breakdown. The drop in pH is thus a key factor influencing the efficiency and speed of decomposition, amplifying the degradation of tissues.

Mechanisms Behind pH Reduction

The decline in pH is primarily caused by the continued buildup of lactic acid, which is produced as a result of anaerobic glucose metabolism. Under normal metabolic conditions, glucose is broken down aerobically, generating more efficient energy sources and fewer acidic byproducts. In contrast, anaerobic metabolism leads to increased production of lactic acid due to its lower efficiency and greater generation of acidic metabolites. Without sufficient oxygen, the body cannot process or neutralize lactic acid, resulting in its accumulation in the tissues.

The sustained accumulation of lactic acid and other acidic byproducts progressively lowers the pH of the tissue. This acidification significantly impacts biochemical processes within the body, accelerating tissue decomposition by enhancing enzymatic activity. The heightened enzymatic activity under these conditions contributes to the rapid breakdown of cellular structures and alters the course of decomposition. The reduction in pH and the associated buildup of acidic metabolites are thus central to shaping the dynamics of postmortem changes.

Significance of pH:

The pH of tissue serves as a critical indicator of its condition and level of decomposition. A low pH signals a high concentration of acidity, which enhances the activity of enzymes responsible for breaking down cellular structures, thereby expediting tissue decomposition. Forensic experts use this information to better determine the timing and circumstances of death. By analyzing the acid-base ratio in various tissue samples, investigators can gain valuable insights into the timeline of postmortem changes. This is particularly useful in cases where the exact time of death is unclear.

A detailed examination of pH levels and the biochemical processes influencing them allows for a more precise reconstruction of the final moments of life and the immediate postmortem changes. Through these analyses, forensic experts can draw more accurate conclusions about the circumstances of death and more effectively narrow down the time of death. A thorough investigation of pH provides essential information for forensic studies and plays a vital role in reconstructing the sequence of postmortem events.

4.3. Microbiological changes

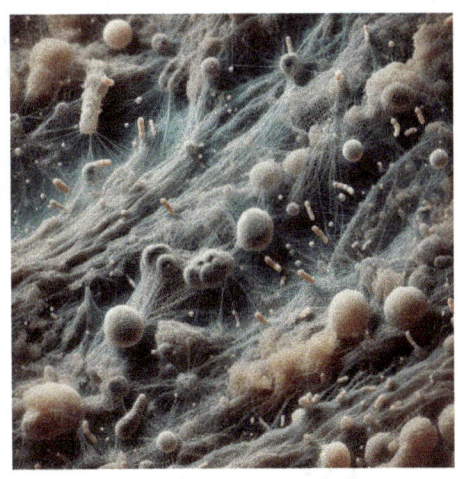

In the initial hours after death, complex microbial processes begin to significantly contribute to the body's decomposition. These changes are driven by the interplay of various microorganisms, such as bacteria and fungi, originating from both the body itself and the surrounding environment. Microbial activity marks the onset of decomposition, triggering profound biochemical reactions that determine the progression of decay. Understanding these processes is critical to comprehending the postmortem changes and offers insights into the mechanisms through which the body undergoes decomposition after death.

4.3.1. Microbial Invasion
Penetration of Microbes
Shortly after death, as the body's physiological functions cease and the immune system becomes inactive, an intense microbial invasion begins. Microbes that previously existed in harmony with the body, such as those in the digestive tract and on the skin, now have the opportunity to spread without restriction. These microorganisms exploit the nutrient-rich environment within the body for growth and reproduction. Without natural defense mechanisms to prevent their spread, bacteria and fungi rapidly invade internal tissues and organs.

This marks the start of a new phase of decomposition, where microbes dominate the biological environment and accelerate tissue breakdown.

Types of Microorganisms Involved

The primary microbes responsible for postmortem decomposition include both aerobic and anaerobic bacteria:

Escherichia coli, commonly found in the human gut, becomes particularly active postmortem due to its ability to degrade various organic compounds, producing byproducts that accelerate decomposition.

Clostridium species, anaerobic bacteria thriving in oxygen-deprived conditions, play a significant role by producing large amounts of gases and acids, contributing to body bloating.

Bacteroides, another major group of bacteria present in the colon, actively break down undigested food and tissues.

Fungi, such as yeasts like *Candida*, also participate in decomposition by degrading organic materials and supporting microbial activity.

Onset of Fermentation

During microbial invasion, the fermentation of organic substances begins. This process involves microbes breaking down carbohydrates, proteins, and lipids, producing a variety of byproducts. These include gases such as carbon dioxide, methane, and hydrogen sulfide, as well as organic acids. Gas accumulation in body cavities and tissues causes visible bloating and swelling. Meanwhile, the organic acids increase tissue acidity, further accelerating decomposition. Fermentation is critical to the rapid breakdown of tissues, intensifying microbial activity and propelling the decay process.

Enzymatic Activity

Microbes release numerous enzymes that significantly speed up tissue degradation:

Proteases break down proteins into smaller peptides and amino acids.
Lipases decompose fats into fatty acids and glycerol.
Cellulases degrade cellulose into sugars.

These enzymes are essential for dismantling complex organic molecules into simpler compounds that microbes use as energy sources. Enzymatic degradation breaks down cellular structures, leading to further tissue destruction and accelerating the body's decay. This enzymatic activity enables microbes to efficiently utilize the remaining organic materials, facilitating comprehensive decomposition.

Influence of Environmental Factors

The speed and extent of microbial invasion are strongly influenced by external factors such as temperature, humidity, and the presence of insects:

Higher temperatures promote microbial growth and activity by accelerating enzymatic reactions and reproduction.
Higher humidity supports bacterial and fungal proliferation by providing a moist environment conducive to microbial activity.
Lower temperatures and **dry conditions**, on the other hand, slow microbial activity by inhibiting their growth and reproduction.
Insects, particularly fly larvae, also influence decomposition by mechanically breaking down tissues, further supporting microbial activity.

The interplay between these environmental factors and microbial processes determines the rate and scope of decomposition. Understanding these interactions is crucial for analyzing postmortem timelines and the biological dynamics of the body after death.

4.3.2. Gas Formation

Generation of Gases

Following death, an extensive microbial process begins, during which various bacteria start decomposing the body's tissues. This microbial activity produces a range of gases as byproducts. The primary gases generated during this stage include methane (CH_4), hydrogen (H_2), and hydrogen sulfide (H_2S). Methane is produced through the anaerobic breakdown of organic compounds by certain bacteria, which proliferate extensively in the intestinal tissues. Hydrogen results from the decomposition of carbohydrates and proteins, while hydrogen sulfide, known for its distinct rotten egg smell, is released during the breakdown of sulfur-containing amino acids. These gases accumulate in body cavities and tissues, significantly contributing to the physical transformations observed in the corpse.

Visible Bloating

The buildup and pressure of these gases result in noticeable bloating, referred to as postmortem emphysema, which is one of the most apparent signs of advanced decomposition. The body, especially the abdomen and other cavities, begins to swell visibly. The skin becomes distended due to the accumulation of gases, leading to a marked increase in tension. In severe cases, the bloating can become so extreme that it causes visible ruptures in the skin. This bloating alters the external appearance of the body significantly, providing investigators with a clear indication of the decomposition stage.

Putrefactive Odor

The gases generated during microbial decomposition not only have physical effects but also contribute significantly to the distinct odor of putrefaction. Hydrogen sulfide, released from the breakdown of sulfur-containing compounds in the body, is particularly notorious for its sharp, unpleasant smell of rotten eggs. Alongside hydrogen sulfide, other volatile organic compounds (VOCs) are produced, intensifying the overall odor. These odors serve as strong indicators of decomposition progression and are often used by forensic experts to estimate the postmortem interval. The specific smell can vary depending on the microbes involved in the decomposition process and the environmental conditions.

Gas Expulsion

As decomposition progresses, the accumulated gases find pathways to escape from the body, typically through natural orifices such as the mouth, nose, and anus. Gas release is often accompanied by the exudation of fluids resulting from tissue breakdown. This process not only alters the body's external appearance but can also impact the surrounding environment. The expulsion of gases can produce a strong, pervasive odor that spreads throughout the area, presenting additional challenges for handling the remains.

Forensic Implications

A detailed examination of gas production and distribution within the body can provide forensic investigators with valuable insights into the timing and progression of decomposition. Forensic experts analyze patterns of gas accumulation and release to estimate the time of death. Deviations in gas distribution or release may also indicate whether the body was moved or manipulated postmortem. Careful analysis of gas-related changes in the body and the surrounding environment can aid in reconstructing the circumstances of death, ensuring the accuracy and integrity of the forensic investigation.

Energy and Transience -The Journey After Death

4.4. Other physiological changes

The initial hours following death are marked by numerous additional physiological changes that are both visually and olfactorily significant. These changes include various skin transformations and the development of characteristic odors associated with the body's advancing decomposition. These processes are critical for forensic analysis as they provide valuable information about the time and conditions of death.

4.4.1. Skin Changes

Lividity Formation

Immediately after death, blood begins to settle into the lower regions of the body due to gravity. This results in a visible discoloration of the skin known as livor mortis or postmortem lividity. This discoloration is particularly evident in areas in contact with the surface on which the body lies. In these regions, the skin appears purplish to dark red and can darken further over time as blood coagulates and pools within the tissues. The intensity and distribution of livor mortis can be influenced by various factors, including ambient temperature, humidity, and the body's position after death. Pressure applied to the body in certain positions can also intensify or diminish these discolorations.

Skin Collapse

As decomposition progresses, the skin gradually loses its elasticity and moisture. This process leads to a visible transformation in skin texture, making it appear wrinkled and slack. These changes are primarily due to the degradation of collagen and elastin within the connective tissues. This degradation typically begins a few hours after death and intensifies as decomposition advances. The absence of moisture further exacerbates the deterioration of the skin, making it more susceptible to tearing and other mechanical damage. External factors such as environmental temperature and humidity can accelerate this process, contributing to the collapsed appearance of the skin.

Blistering and Peeling

With continued decomposition, blisters filled with decomposition fluids may form beneath the skin. These blisters arise from the accumulation of gases and fluids released during the breakdown of tissues. When these blisters burst, the outer layers of the skin may peel away, exposing the underlying tissues. This process, known as skin slippage or exfoliation, can accelerate the visible degradation of the skin. The formation of blisters and subsequent peeling are clear indicators of the intensity of decomposition and can provide forensic experts with valuable insights into the postmortem interval and environmental conditions. Peeling can also be influenced by external mechanical factors, such as movement of the body or contact with external materials.

Additional Changes

Beyond the described skin alterations, other physiological adjustments are also observed. For instance, the skin color can change during the early hours after death, ranging from pale to grayish to greenish, depending on the extent and location of blood pooling and the specific conditions of decomposition. The formation of enzymatic blisters and the release of decomposition-related odors are also characteristic signs of advancing decay that hold forensic significance. All of these changes provide essential information about the state of the body and the environmental conditions surrounding death.

4.4.2. Odor Formation

Decomposition Odor

The decomposition of a body is a highly complex process in which both enzymatic and microbial activities lead to the release of numerous volatile compounds responsible for the characteristic smell of decay. This pungent and pervasive odor is produced by the breakdown of proteins, lipids, and carbohydrates within the body's tissues. The primary contributors to this odor include compounds like cadaverine and putrescine, which are byproducts of protein decomposition. Cadaverine, a diamine, and putrescine, another amine, are released during the breakdown of amino acids and significantly contribute to the distinctive "odor of death." Additionally, various fatty acids and other organic molecules such as butyric acid and propionic acid are generated during the decomposition process. These molecules, derived from the breakdown of fats and other organic materials, intensify and compound the unpleasant smell associated with decay.

Volatile Organic Compounds (VOCs)

During decomposition, numerous volatile organic compounds (VOCs) are emitted, which significantly contribute to the perception of odor. These compounds are highly diverse and include not only cadaverine and putrescine but also a broad range of substances generated during microbial and enzymatic decomposition. The composition of VOCs can vary significantly depending on several factors, such as the deceased's diet, health status at the time of death, and specific environmental conditions surrounding decomposition. For example, different diets or medications taken by the deceased can alter the type and quantity of VOCs released. Environmental factors such as temperature and humidity also influence microbial activity and the speed of decomposition, subsequently affecting the emission of VOCs.

Forensic Applications

The analysis of decomposition odors and the associated VOCs provides forensic experts with critical insights into the state of decomposition and the timeline since death. The intensity and specific composition of these odors can offer clues about the progression of decomposition and its corresponding timeframes. Odor analysis is particularly valuable in cases where visual examination of the body is challenging or impossible due to advanced decay or other external factors. Forensic scientists utilize specialized tools and techniques, such as gas chromatography-mass spectrometry (GC-MS), to precisely determine the composition of volatile compounds. This allows for a more accurate estimation of the postmortem interval. These findings are essential for reconstructing the events surrounding the time of death and the postmortem phase, helping to identify the time and circumstances of death and detect potential manipulations or changes to the body after death.

Summary

The formation and release of decomposition odors are a significant aspect of early postmortem changes, offering valuable insights for forensic science. The complex mixture of volatile organic compounds generated during decomposition provides critical information about the progression of decay and the circumstances of death. A detailed understanding of odor formation and its chemical underpinnings is therefore invaluable for forensic analysis and reconstructing the circumstances surrounding death.

Chapter 5:
The process of decomposition

Energy and Transience -The Journey After Death

5.1. Introduction of decompositionin

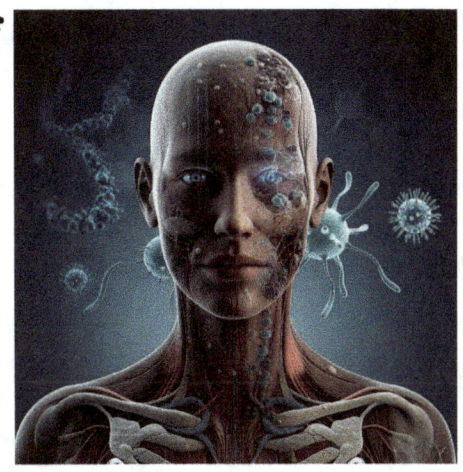

The process of decomposition begins immediately after death and is a complex, multifaceted phenomenon describing the gradual return of body mass to the environment. This continuous process involves a series of biological, chemical, and microbiological changes that collectively work to systematically alter the structure and composition of the body. Decomposition is a natural component of the ecological cycle, necessary for breaking down organic matter into its basic chemical building blocks. This breakdown facilitates the release of nutrients, which can then be reused by other organisms within the ecosystem.

The speed and nature of decomposition are influenced by a variety of internal and external factors. These include ambient temperature, humidity, oxygen availability, the type and number of microorganisms involved, and the presence of scavengers. Each of these factors can uniquely affect the decomposition process, leading to variations in its timeline and patterns.

The interplay of these factors determines the stages of decomposition and their characteristics, marking the transformation from a living to a fully decomposed state. Understanding this process is crucial for ecological studies and forensic investigations, providing insights into the recycling of organic materials and the postmortem timeline.

5.1.1. Definition and Stages of Decomposition

Definition of Decomposition

Decomposition is the biological process in which the complex organic structures of a deceased organism are broken down into simpler chemical compounds through various chemical and microbiological mechanisms. This process is an essential part of the ecological cycle, as it facilitates the return of organic matter to the environment and contributes significantly to nutrient cycling. During decomposition, the complex molecules that make up a living organism's body are reduced to smaller, more stable chemical units. This occurs through the activity of enzymes and microorganisms that degrade cellular structures and tissues. The release of nutrients such as nitrogen, phosphorus, and potassium from decomposed organic material is crucial for the health and growth of plants and other organisms. Decomposition is influenced by various environmental factors, including temperature, humidity, oxygen availability, and the activity of microbes and scavengers. These factors significantly affect the speed and progression of decomposition, resulting in distinct phases and patterns.

Stages of Decomposition

The process of decomposition is divided into several stages, each characterized by specific physical, chemical, and biological changes. These stages are as follows:

Fresh Stage

This stage begins immediately after death and lasts from several hours to a few days. It encompasses the earliest postmortem changes, such as algor mortis (body cooling), livor mortis (postmortem discoloration), and rigor mortis (postmortem rigidity). During this phase, enzymes and microbes begin to break down cellular structures, initiating autolysis. The body starts to cool as metabolic processes that generate heat cease. Blood settles due to gravity in the lower regions of the body, causing the characteristic postmortem discoloration known as lividity. Muscles stiffen and later relax, while initial microbial activities initiate the decomposition process.

Bloating Stage

This stage is marked by the accumulation of decomposition gases such as methane, carbon dioxide, and hydrogen sulfide, which are produced by microbial tissue breakdown. These gases cause a visible and palpable bloating of the body, particularly in the abdominal area. The gas buildup leads to significant internal pressure, distending the skin and underlying tissues. The characteristic odor of decomposition emerges during this stage due to the release of volatile organic compounds (VOCs) such as cadaverine and putrescine. These compounds contribute to the pungent smell associated with advanced decomposition.

Decay Stage

During this stage, microbial activity reaches its peak. The body visibly decomposes as soft tissues are broken down by bacteria, insect larvae, and other scavengers. Liquids released from decomposing tissues further dissolve remaining tissues and create a moist environment conducive to continued microbial activity. This stage is characterized by an intensive breakdown of organic material, with the body progressively disintegrating and collapsing. Tissues and organs begin to dissolve, and the body may be heavily infested with insects and scavengers that accelerate the decomposition process.

Dry Stage

In the final stage of decomposition, primarily bones, hair, and other more resilient structures remain. Microbial activity decreases significantly, and the decomposition process slows down considerably. Over time, even these durable structures are gradually broken down by chemical and physical processes. The dry stage is marked by the reduction of the body to its mineral components. The remaining bones and other hard tissues are eventually eroded by environmental factors such as weathering and chemical degradation. This slow process can take many years and ultimately results in the complete reintegration of the body's remnants into the environment.

5.2. Detailed examination of the decomposition stages

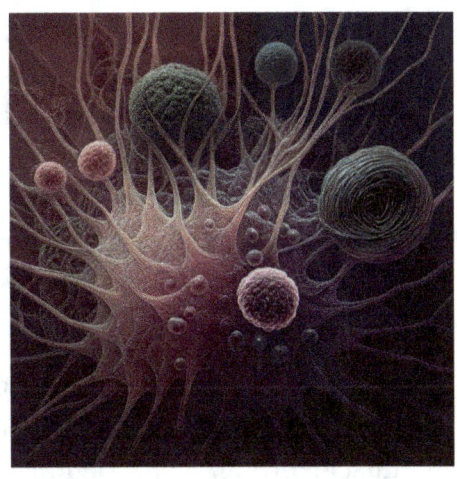

To fully understand the complex process of decomposition, it is essential to examine each stage in detail and analyze the specific biological, chemical, and physical changes involved. Each phase of decomposition is characterized by distinctive features influenced by various internal and external factors. A detailed investigation of these stages provides crucial insights into the dynamics of decomposition and offers valuable information for forensic examinations. Understanding each phase helps to elucidate the underlying processes and their effects on the body.

5.2.1. The Fresh Stage
Onset of Decomposition
Immediately after death, the physiological processes in the body cease. With the halt of blood circulation and oxygen supply, cells stop their metabolism, initiating the decomposition process. At this stage, autolysis begins, wherein enzymes that are normally contained within the lysosomes of cells are released. These enzymes digest the cell membranes and degrade cellular structures. The enzymes involved in autolysis belong to various classes, such as proteases, which break down proteins, and lipases, which decompose fats. The

breakdown begins within cells and gradually extends to adjacent tissues. This process is accelerated by the absence of ATP (adenosine triphosphate), which is essential for maintaining cellular integrity and biological functions. The lack of ATP results in the progressive destruction of cellular structures and functions, ultimately leading to the complete decomposition of cell contents.

Rigor Mortis

Rigor mortis, or postmortem rigidity, begins approximately 2 to 6 hours after death. During this stage, biochemical changes in muscle proteins, particularly actin and myosin, cause the muscles to stiffen. This phenomenon occurs because ATP, which is necessary for muscle contraction and relaxation, is no longer produced. In the absence of ATP, muscles remain in a contracted state, resulting in the characteristic stiffness. Rigor mortis starts in smaller muscle groups, such as those of the eyelids and face, and gradually spreads to larger muscle groups. Maximum stiffness is typically reached within 12 to 24 hours and may persist for up to 48 hours before the muscles begin to relax again. This process is influenced by various factors, including ambient temperature, the overall health of the deceased, and physical activity before death. Higher temperatures accelerate the onset of rigor mortis, while colder temperatures slow it down.

Livor Mortis

Livor mortis, also known as postmortem lividity, occurs when blood settles in the lower parts of the body due to gravity after the heart stops. This results in visible purplish-red discoloration of the skin, which serves as a valuable indicator of the body's position at the time of death. The discoloration typically appears within the first 30 minutes after death and becomes more pronounced over time. Livor mortis can help determine whether the body was moved after death, as the discoloration corresponds to areas where blood has pooled due to gravity. The specific distribution of the discoloration can also provide clues about the position of the body and any potential manipulation postmortem. The analysis of livor mortis is an important forensic tool for determining the time and circumstances of death.

5.2.2. The Bloating Stage

Gas Formation and Swelling

The bloating stage typically begins a few days after death and is marked by significant changes due to the production and accumulation of gases within the body. This occurs primarily due to the activity of microbes, particularly anaerobic bacteria, which break down organic substances in the body. These bacteria thrive in the oxygen-deprived environment that prevails postmortem. The decomposition of carbohydrates, proteins, and lipids by these microbes produces a range of gases, including methane (CH_4), carbon dioxide (CO_2), hydrogen (H_2), and hydrogen sulfide (H_2S). These gases accumulate in body cavities and tissues as a result of enzymatic breakdown of organic matter. Their build-up leads to visible and palpable bloating of the body. The abdomen and other body areas swell significantly, causing a noticeable increase in body size. This bloating is not only a visible indicator of decomposition but also a sign of the advancing microbial activity within the body.

Microbial Activity

During this stage, microbial activity reaches its peak. Anaerobic bacteria proliferate exponentially and produce large quantities of gas as they ferment the remaining organic substances. Gas production leads to increased intra-abdominal pressure, causing tissues to bulge outward and contributing to the body's prominent bloating. The pressure within body cavities continues to rise until it becomes excessive, prompting gases to escape through natural orifices such as the mouth, nose, anus, and other openings. This release of gases is often accompanied by the exudation of fluids, which are also a byproduct of the decomposition process. The combined effect of gases and fluids further degrades the body's external appearance. Microbial activity during this phase is particularly intense, and the progressing decomposition may lead to the formation of blisters on the skin, which can rupture upon contact.

Putrefaction Odor

The characteristic odor of putrefaction, which develops during this stage, is caused by the release of volatile organic compounds (VOCs) produced during microbial decomposition. Among these, hydrogen sulfide (H_2S), known for its distinctive rotten egg smell, as well as cadaverine and putrescine, products of protein breakdown, are particularly prominent. These gases and organic compounds diffuse from the body and spread into the surrounding environment. The odor is strong and can often be detected from a significant distance. It serves as an important indicator of decomposition progression and provides forensic experts with valuable information about the state of the body. Analyzing the putrefaction odor can help assess the extent of decomposition and infer environmental conditions as well as the time elapsed since death. The intensity and composition of the odor can vary depending on the specific microbial communities and environmental factors, offering additional clues about the stage of decomposition.

5.2.3. The Decay Stage
Decomposition of Soft Tissues

During the decay stage, which begins several weeks after death, decomposition processes reach their peak intensity. At this point, the soft tissues of the body are extensively broken down through a combination of microbial activity, insect infestation, and scavenger involvement. Microbes that were active in earlier stages continue their work, breaking down the remaining organic materials into simpler compounds. In this nutrient-rich and significantly altered environment, microbes produce gases and enzymes that further accelerate the breakdown process. Proteins and lipids are enzymatically degraded into amino acids, fatty acids, and other metabolites. Concurrently, insect larvae, often referred to as maggots, feed on the decomposing tissues. These larvae emerge from the eggs laid by flies and other insects, consuming the tissues and introducing additional enzymes and microbes that enhance decomposition. This combination of microbial and insect activity facilitates rapid tissue breakdown.

Insect Colonization

Insects play a pivotal role during the decay stage. Flies, particularly species like the common housefly, are usually the first to detect and colonize the corpse. They deposit their eggs in open body cavities or within the decomposing tissues. The hatched larvae feed voraciously on the decaying material, significantly accelerating decomposition by mechanically breaking down tissues and providing easier access for microbes. This insect colonization is not only a critical component of the decomposition process but also a key factor in the ecological redistribution of nutrients. Insects help transform complex organic matter into simpler substances that can be more readily absorbed by the environment. The enzymes produced by the larvae further expedite tissue breakdown, enhancing the overall efficiency of the decomposition process.

Fluid Seepage

The decay stage is also marked by a significant increase in fluid seepage. These fluids, released as tissues decompose, often accumulate in body cavities and can seep into the surrounding soil or surfaces. Composed of a mixture of decomposed organic matter, microbes, and metabolic byproducts, these fluids can alter the local environment. They enrich the soil with nutrients, facilitating microbial growth and other decomposers. As these fluids percolate through the ground, they can affect the surrounding ecosystem by influencing plant growth or modifying the soil's chemical composition. The seepage is often accompanied by strong odors, which further degrade the surrounding environment by attracting more microbes and insects that continue the decomposition process.

The release of these fluids highlights the interconnected nature of decomposition, where the breakdown of organic matter contributes not only to the decay of the body but also to the transformation of the surrounding environment. This stage underscores the importance of both microbial and insect contributions in recycling organic matter into the ecosystem.

5.2.4. The Dry Stage

Remaining Structures

In the final stage of decomposition, known as the dry stage, most of the body's organic materials have been broken down. What remains are the more resilient structures, such as bones, cartilage, and hair. These robust remnants have proven resistant to the intense decomposition processes that have already disintegrated the soft tissues. Bones, with their dense mineral composition, and cartilage, which is also relatively durable, persist as the last visible remnants of the body. This stage is marked by a significant slowdown in the rate of decomposition, as microbial and enzymatic activity decreases substantially. The remaining structures often exhibit a dry, brittle appearance, reflecting the progressive loss of organic matter and the advanced state of decomposition.

Chemical Breakdown

Even the remaining resilient components, particularly the bones, are not completely immune to degradation. Over time, they undergo further chemical and physical changes that gradually affect their condition. Environmental factors such as soil chemistry, moisture levels, and temperature play a crucial role in the pace and nature of this breakdown. Soil chemistry can influence the dissolution of minerals within the bones, while varying moisture levels may either accelerate or decelerate the chemical degradation process. High moisture levels promote hydrolytic processes that break down bones further, whereas drier conditions tend to stabilize their mineral content. Temperature also has a significant impact, with higher temperatures speeding up the breakdown process and lower temperatures slowing it down.

Ecological Role

The final decomposition of the remaining structures plays a vital role in releasing nutrients back into the soil. These nutrients, previously stored within the body, are broken down into simpler compounds during the advanced stages of decomposition, becoming available to the surrounding environment. Plants can absorb and utilize these nutrients, contributing to soil regeneration and supporting the nutrient cycle within the ecosystem. Microorganisms in the soil also benefit from the released nutrients, continuing the decomposition of other organic material. The gradual degradation of these resilient remnants completes the natural cycle of matter and energy, playing a key role in maintaining ecological balance. This process is not only critical for environmental health but also provides insights into the long-term effects of decomposition on soil chemistry and plant growth.

Energy and Transience -The Journey After Death

5.3. Factors influencing Decomposition

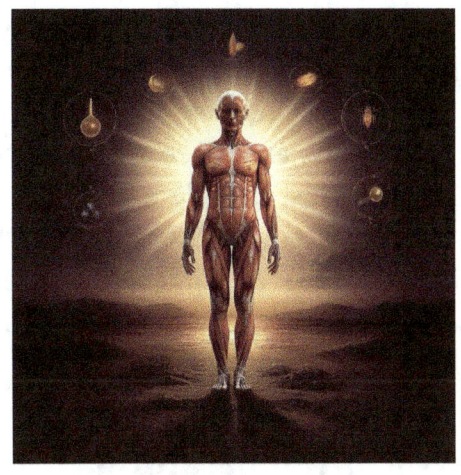

The decomposition of a body is a highly complex process, with its progression and rate heavily influenced by a variety of environmental factors. These factors play a crucial role in either fostering the conditions under which microbial activity thrives or creating barriers that inhibit these processes. A detailed examination of temperature, moisture, and overall environmental conditions is essential to understand the extent and speed of decomposition. These elements interact in a complex interplay, with each component shaping the decomposition process in distinct ways.

5.3.1. Temperature

Temperature is one of the most critical factors affecting decomposition, as it directly impacts microbial activity and the rate at which enzymatic processes occur.

Cold Temperatures

In low-temperature environments, the decomposition process slows down considerably. This deceleration occurs because both microbial activity and enzymatic reactions become significantly less efficient in colder settings. Microorganisms responsible for breaking down tissue are temperature-dependent, and their reproduction rates drop as temperatures fall. Enzymes involved in the degradation of proteins, fats, and carbohydrates also lose efficiency in the cold, reducing the chemical reactions occurring within the body. In extreme cases, such as in freezing conditions, decomposition can come to an almost complete halt. This leads to a significant delay in the decaying process, preserving the body in a near-static state. Such conservation effects are especially evident in cold climates or during winter months, where visible signs of decomposition, such as tissue breakdown and gas formation, are greatly diminished.

Warm Temperatures

Conversely, high temperatures drastically accelerate the decomposition process. Heat acts as a catalyst for enzymatic activity and promotes the reproduction of microbes responsible for breaking down the body. At elevated temperatures, enzymes work faster and more efficiently, leading to the rapid breakdown of tissues. Microorganisms that thrive in warm conditions multiply exponentially, and their heightened activity accelerates the decomposition process. The combination of high temperature and humidity can further intensify this effect, creating optimal conditions for microbial activity. In tropical or subtropical climates, where temperatures remain consistently high, decomposition can advance significantly within just a few days. In such environments, bodies decompose at a noticeably faster rate compared to colder regions, with the progression of tissue decay and other postmortem changes occurring at an accelerated pace.

5.3.2. Moisture

Moisture is another key factor that significantly affects the speed and nature of decomposition. The water content in the environment can either enhance or inhibit microbial activity, depending on whether conditions are wet or dry.

High Moisture

In environments with high humidity or the presence of water, decomposition is significantly accelerated. Moist conditions provide an ideal environment for microorganisms responsible for breaking down the body. Water is essential for microbial activity, as microbes rely on moisture for survival and reproduction. Additionally, water facilitates the diffusion of enzymes and nutrients through the tissues, expediting decomposition. Moist conditions soften cellular structures, making it easier for microorganisms to access nutrients. As a result, the decomposition process progresses rapidly, leading to the liquefaction of tissues and the production of an intensified odor caused by the release of gaseous byproducts of decay.

Dry Conditions

Conversely, dry conditions drastically slow down decomposition. In low-humidity environments or arid regions, the lack of moisture severely restricts the decomposition of the body. Microorganisms require water to survive and maintain their enzymatic processes, and without sufficient moisture, they cannot function effectively. This causes the tissues to dry out, often resulting in a form of natural mummification where the body decomposes much more slowly. The desiccation of tissues inhibits microbial activity since microbes cannot survive without adequate moisture. In such dry environments, decomposition can be delayed for weeks, months, or even years, with the visible signs of decay progressing slowly and the body remaining relatively well-preserved.

5.3.3. Environmental Conditions

In addition to temperature and moisture, other environmental conditions significantly influence the decomposition process. Soil composition and animal activity are critical factors that can affect how a body decomposes.

Soil Type

The soil in which a body rests has a substantial impact on the decomposition process. Different soil types, such as sandy, clayey, or calcareous soils, influence how the body breaks down. Sandy soils tend to drain water quickly, leading to faster desiccation of the body. In contrast, clayey soils retain moisture for longer periods, potentially slowing the decomposition process. Calcareous soils, often alkaline in nature, may further slow decomposition by inhibiting the activity of acid-loving microbes. The chemical composition of the soil affects the microbial communities involved in decomposition and the rate at which the body breaks down. Soil pH, microbial presence, and nutrient availability all play crucial roles in determining the speed of decomposition.

Animal Activity

The presence and activity of animals also play a significant role in the decomposition process. Animals such as insects, rodents, and other scavengers accelerate decomposition by consuming the body and breaking down tissues. Insects, especially flies, lay their eggs in body openings or on the skin, and the resulting larvae (maggots) feed on decaying tissues. This activity speeds up decomposition as larvae not only mechanically disrupt tissues but also release enzymes that promote decay. Larger scavengers such as rodents or wild animals also contribute by consuming parts of the body, hastening tissue breakdown. These animals can pierce thicker layers of skin, which facilitates decomposition by granting microorganisms access to deeper tissues. Animal activity is particularly significant in open environments where the body is easily accessible, allowing animals to feed uninterrupted.

By influencing the rate and manner of tissue breakdown, moisture, soil type, and animal activity shape the dynamics of decomposition, providing valuable insights into the environmental and ecological context of the decaying process.

5.4. Microbiological Aspects of Decomposition

The microbiological aspects of decomposition are essential to understanding this intricate and multifaceted process. Microorganisms, particularly various bacterial species, play a critical role in breaking down organic materials and are instrumental in producing decomposition gases and other chemical compounds. Microbial activity drives the progression through the different stages of decomposition and has a profound impact on the overall process.

5.4.1. Microbes and Bacteria
Role of Bacteria
Bacteria are the primary agents responsible for the decomposition of organic matter, initiating the process immediately after death. The bacterial population within the body is highly diverse and includes numerous species that become active during different phases of decomposition. Early in the process, when oxygen is still available, aerobic bacteria dominate. These oxygen-dependent bacteria begin breaking down complex organic molecules into simpler compounds. As the oxygen levels within the body diminish and anaerobic conditions prevail, anaerobic bacteria take over. These bacteria thrive without oxygen and produce fermentation gases, such as

methane and hydrogen sulfide, as byproducts of their metabolic processes. The transition from aerobic to anaerobic decomposition represents a critical shift in the decay process. Bacterial activity triggers a cascade of biochemical reactions that accelerate the breakdown of proteins, lipids, and carbohydrates, contributing to the comprehensive decomposition of the body. These bacteria produce a wide range of enzymes, including proteases, lipases, and nucleases, which target specific cellular components, facilitating the rapid disintegration of tissues.

Impact of Microbes

Microbes from various sources significantly contribute to decomposition. Primary participants include bacteria naturally present in the gastrointestinal tract. These bacteria begin to break down the intestinal walls shortly after death and quickly spread to surrounding tissues. In addition to gut bacteria, microbes residing on the skin and in the surrounding environment play a pivotal role. Skin and environmental microbes enter the body through openings such as the mouth, nose, wounds, or lesions and further assist in the breakdown process.

These microbes produce an array of enzymes that dismantle chemical bonds in organic matter, leading to the formation of decomposition products like ammonia, sulfur compounds, and organic acids. These compounds are not only responsible for the characteristic odor of decay but also drive further chemical and microbial degradation of tissues. Microbial decomposition is a dynamic process that continuously attracts new microbial populations, influencing the overall pattern and progression of decay.

Microbial Interactions

During decomposition, various microbial species interact and influence one another. These interactions can be either synergistic or antagonistic, affecting the speed and pattern of decomposition. Synergistic interactions occur when multiple microbial species collaborate to optimize a shared metabolic process. Conversely, antagonistic effects arise when one microbial species inhibits the activity of another. These complex interactions contribute to the variability in decomposition rates and patterns across different environments.

Factors Influencing Microbial Decomposition:
Microbial activity is governed by several factors, including temperature, moisture, and pH levels.

> ***Temperature:*** High temperatures enhance microbial growth and activity, accelerating decomposition, while lower temperatures slow this process.
> ***Moisture:*** Adequate moisture supports microbial survival and reproduction, facilitating enzymatic breakdown of tissues.
> ***pH Levels:*** A balanced pH is crucial for microbial activity, as extreme pH levels can inhibit microbial efficiency.

By driving enzymatic activity, gas production, and tissue breakdown, microbes form the backbone of the decomposition process. Understanding their roles and the factors that influence their activity provides invaluable insights into the complexities of decay and the ecological interactions surrounding it.

5.4.2. Microbiomes and Decomposition
Microbiomes in the Body
The microbiome of a body, comprising the entirety of microorganisms living on and within it, exerts a profound influence on the decomposition process. Every individual possesses a unique microbiome shaped by various factors, including lifestyle, diet, health, and environmental conditions. These microbial communities play a crucial role in determining the speed and pattern of decomposition. Different

microbial species bring distinct enzymes and metabolic pathways, each targeting specific body components. For instance:

- Certain bacterial species focus on breaking down proteins by producing specialized proteases that degrade large protein complexes into smaller peptides and amino acids.
- Others specialize in lipid degradation, employing lipases to break fats into fatty acids and glycerol.
- Additional microbes target carbohydrates, using enzymes like amylases to decompose these compounds.

The interaction and coordination of these microbial activities significantly influence the pace and manner of decomposition. Thus, the microbiome is not merely a passive participant but an active driver of the process, substantially shaping the progression of decomposition through its composition and activity.

Interactions Between Microbial Species

Interactions among microbial species are central to the decomposition process. Microbes often work synergistically, where the metabolic byproducts of one species serve as substrates for others. This complex network of interactions ensures coordinated decomposition and largely dictates the enzymatic activities necessary for breaking down organic materials.

For example, volatile compounds produced by aerobic bacteria during the early stages of decomposition can alter conditions—such as pH and oxygen levels—to favor anaerobic bacteria. These anaerobes then take over in the later stages, producing fermentation gases like methane and hydrogen sulfide. The efficient collaboration among microbial species is vital for successful decomposition, as one species' byproducts often create the conditions necessary for the growth and activity of others.

These interactions illustrate the dynamic equilibrium within the microbiome, which governs the different phases of decomposition. Microbial activity is influenced not only by the enzymatic capabilities of individual species but also by environmental conditions shaping the microbiome's function.

Impact of Environmental Factors

The conditions within the microbiome are affected by various environmental factors that influence microbial activity and, consequently, the rate of decomposition. Key factors include:

> ***Temperature:*** High temperatures accelerate microbial activity, enhancing decomposition rates, whereas low temperatures slow these processes considerably.
>
> ***Moisture:*** Microbes rely on water for growth and reproduction. Optimal moisture levels promote microbial activity and support decomposition, while overly dry or excessively wet conditions may inhibit the process.
>
> ***pH Levels:*** Microbial enzyme activity is sensitive to pH levels, with extreme pH conditions disrupting enzymatic processes and hindering microbial decomposition.

These factors collectively shape the microbiome's characteristics and its capacity to drive decomposition. The intricate balance between these elements determines the effectiveness and speed of microbial activities during the decomposition process.

Summary

The decomposition process is a complex, dynamic sequence that begins immediately after death and is characterized by a combination of biological, chemical, and microbiological factors. The early stages involve autolysis and microbial activity, followed by intense gas production and tissue breakdown during the bloating phase. During the active decay phase, most soft tissues are degraded, leaving resilient structures like bones. Environmental factors, such as temperature, moisture, and soil composition, play critical roles in influencing the speed and nature of decomposition.

At the heart of this process lies the microbiome, comprising a diverse array of microorganisms, primarily bacteria, which drive and regulate decomposition. Interactions among microbial species and their adaptation to environmental conditions critically determine the progression and efficiency of decomposition. Understanding these processes is vital for forensic science and ecological research, offering insights into postmortem changes and their implications for the decomposition process.

Chapter 6:
Long-term energy Transformations

After death and the onset of decomposition, a lengthy and intricate process of energy transformation unfolds, extending far beyond the immediate changes within the body. These long-term energy transformations are crucial for understanding the finite cycle of matter and energy in nature. Following death, the body undergoes various stages in which the energy stored in organic substances is converted into other forms and subsequently integrated into the ecological system. This process encompasses multiple steps, starting with the decomposition of organic molecules and the breakdown of body mass and culminating in the eventual return of energy and nutrients to the soil, where they can be reabsorbed by plants and other organisms. Through these continuous transformations, energy is ultimately reintegrated into the ecological cycle, which serves as the foundation of life on Earth.

6.1. Energy Transformations after Decomposition

6.1.1. Conversion of Organic Molecules

Decomposition of Organic Substances

Following the death of an organism, a highly intricate and multi-faceted process of decomposition begins, primarily driven by microbial activities, including those of bacteria and fungi, as well as enzymatic processes. The large, complex organic molecules that constitute the body—proteins, lipids, and carbohydrates—are broken down in a series of steps. This begins immediately after death, as cellular metabolic processes cease. The breakdown of these molecules is facilitated by various microbial and enzymatic activities.

Proteins, the body's fundamental building blocks, undergo a process called proteolysis, during which they are broken down into their smaller components, amino acids. This process is vital for further decomposition, as amino acids are more easily utilized by many microbes. These amino acids are then further processed by various microbes, either dismantling them into simpler components or integrating them into their own metabolic pathways.

Lipids, primarily serving as energy reserves within the body, are broken down into glycerol and fatty acids by lipases—

enzymes specialized for fat decomposition. Microbes utilize these fatty acids for energy production or further reduce them into smaller molecules that can also be used by other organisms. Lipid breakdown plays a particularly important role in fueling the microbes that drive the decomposition process.

Carbohydrates, present as polysaccharides like glycogen and cellulose, are degraded by enzymes such as amylases and cellulases into simpler sugars like glucose. Microbes rapidly metabolize these sugars, further transforming the energy stored within them. The conversion of carbohydrates to glucose is a swift process, providing microbes with immediately available energy.

Byproducts of Decomposition

The byproducts generated during the decomposition process are predominantly water-soluble and can be readily absorbed by microorganisms, fungi, and other soil-dwelling organisms. These byproducts, including amino acids, simple sugars, and fatty acids, retain substantial amounts of energy, which are critical for microbial life. These organisms harness the released energy to sustain their biological processes and growth, thereby perpetuating a continuous cycle of energy transformation.

Fungi and bacteria play a pivotal role in this process, as they are responsible for breaking down organic material and making its nutrients available once more. The decomposition of organic material releases nutrients into the soil, thereby preserving soil fertility. These nutrients are subsequently absorbed by plants, which utilize them for growth. This mechanism returns the energy from decomposition byproducts to the biological cycle, thereby closing the natural loop of matter and energy.

In summary, the long-term energy transformations following death constitute a complex and essential process that involves the breakdown of organic substances into simpler molecules and their integration into the ecological system. These processes are fundamental to understanding ecological cycles and maintaining soil fertility, which is vital for plant growth and the support of life on Earth. A comprehensive understanding of these energy transformations enhances our knowledge of ecological processes and the natural cycle of matter and energy.

6.1.2. Mineralization
Transformation into Minerals

One of the most significant aspects of long-term energy transformations following the death of an organism is the process of mineralization. Mineralization refers to the conversion of organic substances into mineral components, which are more stable and can remain in the soil over extended periods. This process is vital for the ecological cycle as it ensures the return of nutrients to the soil, where they can be absorbed by plants. Mineralization occurs through a combination of microbial activity and chemical processes that work together to transform organic materials into inorganic minerals. These minerals include phosphates, carbonates, sulfates, and other compounds essential to ecological balance.

During mineralization, organic phosphates present in residual tissues and body fluids are broken down into inorganic phosphates through microbial decomposition and chemical reactions. These inorganic phosphates are crucial for plant growth, as they are integral to plant nutrient systems. The transformation of organic phosphates into inorganic forms ensures their availability for plants, which is critical for maintaining soil fertility and supporting new plant growth.

Another vital process within mineralization is the transformation of organic nitrogen compounds into inorganic forms such as nitrates and nitrites through nitrification and denitrification. These processes are facilitated by specialized microbes capable of breaking down nitrogen compounds in organic material and converting them into forms usable by plants. Nitrates and nitrites are essential components of the nitrogen cycle and play a critical role in synthesizing amino acids and other biological molecules in plants.

Soil Formation

The mineral components produced during mineralization significantly contribute to soil formation. The availability of minerals in the soil is crucial for plant nutrition and soil fertility. Mineralization provides a continuous source of nutrients that plants can absorb through their roots, closing the nutrient flow cycle. These nutrients, supplied through decomposition and mineralization, directly support plant growth and development.

Mineral components not only enhance nutrient availability but also influence the physical properties of the soil. Minerals contribute to aggregate formation, where smaller soil particles bind together into larger aggregates, improving soil structure. Good soil structure is essential as it increases the soil's water-holding capacity, making water more readily available to plants over extended periods. This improvement in soil structure is particularly important for agricultural productivity and the preservation of natural ecosystems, as it enhances soil quality for crop growth and ecological stability.

Moreover, the minerals resulting from mineralization affect the soil's chemical properties, such as pH levels, which are critical for nutrient availability to plants. A well-balanced pH ensures that nutrients remain in forms easily accessible to plants, promoting long-term soil fertility and stability. This is fundamental for sustainable agricultural practices and the conservation of natural ecosystems.

Summary

In summary, the process of mineralization is a crucial phase of long-term energy transformations after death. By converting organic substances into stable mineral components, the energy stored in the body is transformed into a form accessible to other organisms. This process is central to the natural cycle of energy and matter, ensuring that the energy once stored in a living being is not lost but reintegrated into the biosphere to support new life processes. Mineralization helps maintain soil fertility and ecological balance by providing nutrients essential for plant growth and ecosystem functionality. It highlights the interconnectedness of biological processes and the enduring legacy of life within the natural cycle of matter and energy.

Energy and Transience -The Journey After Death

6.2. Impact on the Ecological System

6.2.1. Nutrient Cycles

Carbon Cycle

The carbon cycle is a fundamental component of global ecology, describing the continuous exchange of carbon compounds between the atmosphere, biosphere, and lithosphere. This cycle begins with the decomposition of organic material, during which complex carbon compounds are released. Following the death of an organism and the initiation of decomposition, carbon compounds are released into the atmosphere in the form of carbon dioxide (CO_2). Carbon dioxide is primarily produced by the respiration of decomposer organisms, as well as by the microbial and chemical breakdown of organic material.

This carbon dioxide is then absorbed by plants through the process of photosynthesis. During photosynthesis, plants use light energy to combine carbon dioxide with water, synthesizing organic compounds such as glucose. This biomass forms the foundation of the food chain, as plants are consumed by animals. The organic compounds within the biomass are further broken down into smaller molecules through the digestive processes of animals and microorganisms. Through this cyclical process, carbon compounds return to the soil, where they are either released back into the atmosphere or stored in the soil as organic substances. The reintegration of carbon into the soil also occurs during the decomposition of plant material and animal remains.

The ongoing decomposition and associated carbon exchange contribute to maintaining the carbon balance in both the atmosphere and the soil. This balance is critical for the global carbon budget and has profound effects on the climate and ecological equilibrium. Disruptions to this cycle, such as those caused by deforestation or intensive agriculture, can lead to increased carbon dioxide concentrations in the atmosphere, which, in turn, accelerates climate change.

Nitrogen Cycle

The nitrogen cycle is another essential component of the ecological system, detailing the transformation and exchange of nitrogen compounds between the atmosphere, biosphere, and soil. Nitrogen is a critical element for plant growth, as it is a key component of amino acids, proteins, and nucleic acids. The nitrogen cycle begins with the decomposition of organic material, during which nitrogen compounds such as proteins and amino acids are broken down into simpler forms, such as ammonia (NH_3).

This process is initiated through microbial decomposition, particularly by ammonification bacteria, which convert nitrogen-containing organic materials into ammonia. Ammonia is then converted into nitrite (NO_2^-) and subsequently into nitrate (NO_3^-) through a microbial process known as nitrification. Nitrate is a form of nitrogen that plants can readily absorb. Plants utilize this nitrate to synthesize proteins and other essential biomolecules.

Once used by plants, nitrogen can return to the soil when plant material and animal excrement decompose. The breakdown of plant residues and animal remains containing nitrogen compounds ensures the continuous replenishment of nitrogen in the soil. This mineralization of nitrogen compounds ensures a steady supply of nitrogen-rich nutrients necessary for plant growth. Additionally, denitrification occurs, during which excess nitrate is converted into nitrogen gas (N_2) by specific bacterial species and released back into the atmosphere.

This cyclical process ensures that nitrogen compounds continuously circulate between the soil and atmosphere, providing the ecosystem with essential nutrients. Disruptions to the nitrogen cycle, such as those caused by excessive fertilization or environmental pollution, can lead to imbalances that reduce soil fertility and destabilize the ecological equilibrium.

6.2.2. Energy Flows in Ecosystems

Energy Transfer

The decomposition of organic material serves as both a mechanism for nutrient recycling and a critical process for energy transfer within ecosystems. When organic materials break down, the energy stored in their molecular structures is released and absorbed by decomposer organisms such as insects, worms, and various soil-dwelling species. These detritivores feed on decaying organic matter, metabolizing the energy for their own survival and biological functions. By doing so, they play a pivotal role in introducing this energy into the broader ecological system.

Through their metabolic processes, detritivores break down complex organic compounds into simpler forms, which other organisms can utilize. The energy absorbed by these primary decomposers does not remain static; instead, it moves up the trophic levels. When detritivores are consumed by larger animals, such as birds, mammals, or other predators, the energy stored in their bodies is transferred to these consumers. This cascading energy transfer ensures a continuous circulation of energy throughout the ecosystem, linking primary decomposers to secondary and tertiary consumers.

These flows are vital for maintaining the integrity and functionality of ecosystems. They create a dynamic cycle where the energy captured by primary producers (plants) through photosynthesis is not wasted but systematically redistributed across various trophic levels. Each organism that participates in this cycle contributes to sustaining biodiversity, promoting ecosystem balance, and ensuring that energy remains available to support life processes.

Ecological Significance

The systematic decomposition of organic material and the associated transfer of energy are indispensable for the health and stability of ecosystems. These processes ensure the continual recycling of resources, preventing the depletion of essential nutrients and energy. By reintroducing these elements into the ecosystem, decomposition processes support soil fertility, plant growth, and the overall productivity of ecosystems.

The role of decomposers extends beyond simply breaking down organic matter; they facilitate nutrient cycling that directly impacts habitats and the availability of resources for all organisms within the ecosystem. As energy flows from one trophic level to the next, it fuels the biological processes of a wide variety of species, contributing to the richness and diversity of life within an ecosystem. This continuous circulation of energy and nutrients underpins the long-term sustainability of ecosystems, ensuring that life can thrive even as individual organisms perish.

Moreover, the efficiency of energy transfer and nutrient recycling determines an ecosystem's resilience to environmental changes and disturbances. Systems with robust decomposition and nutrient cycling processes are better equipped to recover from disruptions, such as natural disasters or human activities, maintaining their ecological balance and functionality.

In summary, decomposition and energy flows within ecosystems are fundamental to their health and sustainability. These processes enable the seamless recycling of nutrients and energy, providing the foundation for plant growth, supporting biodiversity, and fostering the stability needed for ecosystems to thrive. The interplay between energy transfer and nutrient recycling highlights the interconnectedness of all life forms and underscores the importance of preserving the delicate balance of natural systems.

Energy and Transience - The Journey After Death

6.3. Long-Term Effects on Soil

6.3.1. Soil Improvement

Nutrient Enrichment

The processes of decomposition and mineralization play a vital role in enriching soil with essential nutrients. Following the death and subsequent decomposition of an organism, the nutrients contained within the organic material—such as carbon, nitrogen, phosphorus, and other minerals—are transformed into inorganic forms. This transformation makes nutrients more readily available for plant uptake. The breakdown of organic material by microbes and other decomposers releases these nutrients into the soil, where they are absorbed and utilized by plants.

This continuous enrichment of the soil significantly enhances soil fertility, promoting the growth and development of plants. Healthy plant growth is foundational to the food chain, serving as a source of sustenance for animals and microorganisms alike. Plants not only support the ecosystem by transferring energy and nutrients through trophic levels but also contribute to the stability and health of the environment. They play a key role in maintaining ecological balance by stabilizing the soil and serving as carbon sinks.

Furthermore, nutrient-rich and well-aerated soils provide an optimal environment for the proliferation of microorganisms and insects. These organisms further decompose organic substances and facilitate the release of nutrients, ensuring a steady supply of essential elements. This symbiotic relationship among soil microbes, plants, and other organisms ensures the sustainability of soil fertility, allowing ecosystems to thrive over the long term.

Soil Structure

In addition to nutrient enrichment, the decomposition of organic matter has a profound effect on soil structure. The process leads to the formation of humus, a stable organic compound that plays a pivotal role in enhancing soil quality. Humus is the product of microbial activity and the chemical breakdown of organic material, known for its exceptional ability to retain water and nutrients.

The presence of humus greatly improves the soil's water-holding capacity, enabling it to store moisture and provide a consistent water supply to plants over extended periods. This is particularly important in regions with irregular rainfall or during periods of drought. Humus also enhances soil aeration and drainage, promoting root growth and reducing the risks of waterlogging and root rot.

A soil structure enriched with humus is less prone to erosion. Stable soils with high humus content exhibit greater cohesion, reducing susceptibility to wind or water erosion. This not only preserves the critical topsoil layer necessary for plant growth but also prevents the loss of fertile land.

Moreover, humus contributes to the long-term storage of carbon in the soil. Carbon sequestration in the form of humus helps mitigate climate change by reducing the concentration of greenhouse gases in the atmosphere. This long-term carbon storage is a crucial aspect of the global carbon cycle and plays a significant role in climate regulation.

Summary

The nutrient enrichment and structural improvements brought about by decomposition processes are essential for maintaining soil health and productivity. By enriching the soil with nutrients and enhancing its physical properties, decomposition supports plant growth, strengthens ecological resilience, and reduces environmental degradation. The formation of humus not only improves soil functionality but also contributes to carbon storage, aligning soil health with global climate goals. Together, these processes highlight the critical importance of decomposition in sustaining ecosystems and promoting environmental stability.

6.3.2. Soil Chemical Changes

pH Changes

The decomposition of organic material can significantly alter the soil's pH, a critical factor for nutrient availability to plants. Acid-forming processes, such as the breakdown of organic acids, can lower the soil pH, leading to acidification. Acidic soils often reduce the availability of essential nutrients like phosphorus and micronutrients, as these elements tend to become less accessible in low-pH environments. Acidification can thus impact plant growth negatively by restricting nutrient uptake.

Conversely, alkaline processes, such as the decomposition of protein-rich materials, can raise the soil pH. These alkaline conditions can enhance the availability of certain nutrients and make the soil less acidic. Such pH shifts have a direct influence on soil fertility, shaping the balance of nutrients accessible to plants. Maintaining a balanced soil pH is crucial for optimal plant health, as it ensures effective nutrient absorption and supports robust plant development.

The dynamic nature of soil pH alterations caused by decomposition highlights the need for monitoring and managing soil conditions. By understanding these pH shifts, it is possible to implement strategies, such as soil amendments, to maintain an environment conducive to sustainable plant growth.

Trace Elements and Minerals

The mineralization of organic material during decomposition releases various trace elements and minerals essential for plant growth and soil fertility. Trace elements like iron, manganese, zinc, and copper are required in small quantities for critical physiological processes in plants. These elements, liberated during decomposition, are vital for supporting enzymatic activities and synthesizing chlorophyll, which is necessary for photosynthesis.

The availability of these trace elements in the soil directly affects plant health. Deficiencies in specific micronutrients can lead to visible symptoms of malnutrition, such as stunted growth or discoloration of leaves, ultimately reducing crop yields and overall plant vitality. Through the ongoing process of decomposition and mineralization, these crucial elements are replenished in the soil, ensuring they remain accessible to plants over time.

Supporting Soil Fertility

The chemical transformations that occur in the soil through decomposition and mineralization significantly contribute to maintaining soil fertility and ecological health. By continuously supplying the soil with essential nutrients and trace elements, decomposition ensures that the soil remains productive and capable of supporting robust plant growth. This, in turn, sustains the broader ecological balance and enhances the productivity of ecosystems.

Through these processes, the soil evolves into a nutrient-rich medium capable of supporting diverse plant and microbial communities. The interplay between pH changes and the availability of trace elements underscores the complexity and importance of soil chemistry in driving the long-term sustainability of agricultural and natural ecosystems. By fostering these chemical changes, the decomposition process becomes an integral component of ecological resilience and soil productivity.

Energy and Transience -The Journey After Death

6.4. Long-Term Energy Transformations in Nature

6.4.1. The Formation and Role of Humus

Formation of Humus

Humus is a complex amalgamation of organic compounds resulting from the decomposition of plant and animal matter within the soil. The creation of humus begins as organic material, such as decaying plant residues, animal waste, or other biodegradable substances, enters the soil environment. This material undergoes a transformative process driven by microorganisms, including bacteria, fungi, and actinobacteria, which play a crucial role in breaking down these substances. These microbes secrete enzymes that target complex molecules, such as lignin, cellulose, and hemicellulose, breaking them into smaller, more manageable components.

The degradation of lignin—a highly durable component of plant cell walls—is particularly significant. Lignin is known for its resistance to decomposition, yet it contributes substantially to the formation of stable humic substances, which enhance soil health and fertility. Simultaneously, other energy-rich compounds, such as cellulose, are dismantled and transformed into simpler compounds that contribute to the development of humus. These processes collectively lead to the production of humic substances, which are notable for their stability and their ability to retain nutrients effectively.

Humus formation is not merely the result of decomposition but involves a series of intricate biological and chemical transformations. Over time, these processes yield a highly stable substance that integrates seamlessly into the soil. The resulting humus has exceptional water and nutrient storage capabilities, which are crucial for sustaining plant life. This process can span years or even decades, making humus a long-term reservoir of essential nutrients such as nitrogen, phosphorus, potassium, and trace elements. These nutrients, stored within the humus, are slowly released and made available for plant uptake, fostering ecological balance and supporting robust plant ecosystems.

Humus also serves as an indicator of soil health, fertility, and sustainability. Soils with a higher humus content are often more productive and resilient. Beyond storing nutrients and energy, humus also acts as a critical player in the carbon cycle by sequestering carbon in a stable form, helping to mitigate climate change and preserve ecosystem stability.

Functions of Humus

The functions of humus extend far beyond nutrient storage; it plays an integral role in enhancing the physical, chemical, and biological characteristics of soil. One of its primary contributions is the improvement of soil structure. Humus enhances the water-holding capacity of soil, acting like a sponge that absorbs significant quantities of water and gradually releases it to plants during dry periods. This capacity to retain water is particularly beneficial in arid regions or during drought conditions, where water availability becomes a critical limiting factor for plant survival.

Humus also strengthens soil stability by encouraging the formation of soil aggregates. These aggregates bind soil particles together, reducing susceptibility to erosion by wind or water. This aggregation not only protects the fertile topsoil but also promotes better aeration and drainage, enabling plant roots to grow more effectively. These structural benefits make soil more resilient to external pressures, ensuring long-term agricultural and ecological productivity.

Another critical role of humus lies in its buffering capacity, which helps stabilize soil pH levels. Maintaining an optimal pH range is essential for the availability of nutrients to plants, as extreme acidity or alkalinity can inhibit nutrient uptake. Humus mitigates these conditions by neutralizing excess acids or bases, creating a more hospitable environment for plant growth and microbial activity. This pH stabilization fosters a balanced and thriving soil ecosystem.

Furthermore, humus promotes biological activity within the soil. It serves as a habitat and nutrient source for a wide array of microorganisms that continue to break down organic material, recycle nutrients, and improve soil health. These microbes are vital for maintaining a continuous cycle of decomposition, nutrient availability, and absorption, ensuring that ecosystems remain productive and balanced over time.

Conclusion

In sum, humus is a cornerstone of healthy and fertile soils, with functions that span nutrient storage, water retention, structural improvement, and pH stabilization. By fostering robust microbial communities and enhancing soil resilience, humus supports the long-term sustainability of agricultural systems and natural ecosystems alike. Its formation and role in soil ecosystems exemplify the intricate interplay between decomposition, nutrient cycling, and ecological balance. Through its multifaceted contributions, humus ensures the longevity and vitality of the soil, forming the foundation of life for countless terrestrial organisms.

6.4.2. Impact on Biological Activity
Microbial Populations

The long-term transformation of energy through the decomposition of organic matter in the soil significantly influences the composition and activity of microbial populations. Microorganisms, including bacteria, fungi, actinobacteria, and other microscopic organisms, are indispensable agents in nutrient cycling. They play a central role in ecological systems by breaking down organic material and making the released nutrients available to plants and other soil organisms. This decomposition process is not merely about breaking down matter but also about converting complex organic molecules into simpler compounds that can be more readily absorbed by soil flora and fauna.

The diversity and activity of microbial populations are closely linked to the availability and type of organic materials introduced into the soil. A rich supply of organic matter, such as decayed plant residues, animal excretions, and other organic wastes, creates an optimal environment for a diverse microbial community. This diversity fosters intense microbial activity, as different microbes specialize in processing specific types of organic material and perform varied decomposition processes. For instance, certain bacterial species excel at breaking down cellulose, while others are adept at decomposing lignin. The interaction among these diverse microbial species leads to efficient nutrient release, which improves soil quality and fertility.

Microbial activity in soil is influenced by numerous factors, including temperature, moisture, and pH levels. Warm, moist conditions enhance microbial activity and decomposition, while drought or cold can slow these processes. Certain agricultural practices, such as fertilizers or pesticides, also impact soil microbial communities. Maintaining balanced microbial species and optimal activity is essential for soil fertility and sustainable soil use.

Soil Organisms

In addition to microbes, larger soil organisms such as earthworms, insects, and other detritivores play a crucial role in energy transformation and the decomposition of organic matter. These soil organisms contribute significantly to decomposition by physically fragmenting and digesting organic substances. They break down plant residues, animal remains, and other organic wastes, thereby increasing the surface area available for microbial decomposition. This physical fragmentation substantially accelerates the rate of decomposition, as microbes can more easily access and break down the smaller pieces of organic material.

Earthworms, in particular, are exemplary contributors to both decomposition and soil structure enhancement. Their activities not only facilitate the breakdown of organic matter but also improve the mixing of organic material with the mineral components of the soil, promoting the formation of fertile soil. By burrowing and tunneling, earthworms increase soil aeration and permeability, which supports root growth and reduces the risk of erosion. These activities stabilize and enhance soil structure, making it more suitable for agricultural use and the preservation of natural ecosystems.

The biological processes supported by these soil organisms are vital for the continuous renewal and maintenance of soil fertility. They enhance nutrient availability and reinforce the structural integrity of the soil, creating a foundation for a healthy and productive soil ecosystem. Soil rich in microbial life and soil organisms can perform a wide range of ecological functions, from water and nutrient storage to promoting plant growth and supporting biodiversity.

Summary

The long-term transformation of energy following the death of an organism has far-reaching effects on biological activity within the soil. Microbial populations and soil organisms are central to the decomposition and transformation of organic materials, releasing nutrients and improving soil quality. The diversity and activity of these microbes are closely tied to the availability of organic materials and environmental conditions. Meanwhile, soil organisms such as earthworms contribute to the physical fragmentation and mixing of organic matter, accelerating decomposition and enhancing soil structure. Understanding these processes is critical for the sustainable use and preservation of soils and for fostering a healthy and productive environment.

Chapter 7:
Ecological and physical perspectives

Decomposition and the associated energy transformations are not solely biological processes; they are deeply embedded within ecological and physical systems. This chapter examines the effects of decomposition on ecological structures and the physical processes that accompany these changes. It provides a comprehensive overview of the interactions between decomposition, ecosystems, and the environmental physical conditions.

Decomposition is a fundamental process within the ecological cycle, involving the breakdown of organic substances by microorganisms and other decomposers. This process releases nutrients that plants can reabsorb and transforms energy that is utilized by decomposing organisms. However, decomposition does not occur in isolation. It operates within a complex network of interactions that influence both the structure and function of ecosystems.

Ecological Interconnections

From an ecological perspective, decomposition plays a crucial role in nutrient cycling and ecosystem dynamics. Decomposers break down complex organic materials into simpler compounds, releasing essential nutrients like nitrogen, phosphorus, and potassium back into the soil. These nutrients are then absorbed by plants, forming the foundation of terrestrial food webs. In this way, decomposition sustains primary productivity and supports the broader ecological community.

The carbon cycle exemplifies the ecological significance of decomposition. Organic matter, initially sequestered by plants through photosynthesis, is released back into the atmosphere as carbon dioxide during decomposition. This release is a critical component of Earth's climate regulation, as carbon dioxide functions as a greenhouse gas. Simultaneously, a portion of the carbon is stored in the soil as humus, enhancing soil fertility and contributing to carbon sequestration. This balance between carbon release and storage is vital for maintaining both soil health and climate stability.

Physical Impacts of Decomposition

On the physical level, decomposition shapes landscapes and contributes to soil formation. The chemical compounds released during decomposition interact with soil components, altering soil structure and fertility and promoting humus formation. These physical changes directly impact the plant communities and animal populations that inhabit these soils.

Heat generation is another significant physical phenomenon linked to decomposition. Microbial activity during decomposition produces heat, which can be observed in processes like composting. This heat production affects soil temperature, which in turn influences other soil processes and organisms. For instance, elevated soil temperatures can accelerate microbial activity, creating a feedback loop that enhances decomposition.

Environmental conditions such as temperature, moisture, and oxygen availability greatly influence the rate and efficiency of decomposition. Warm, moist environments tend to accelerate decomposition, while colder or drier conditions slow the process. This variability means that the ecological effects of decomposition differ across climates and microclimates. For example, tropical regions experience rapid nutrient cycling due to high decomposition rates, while boreal and arid regions may see slower organic matter turnover.

Complex Interactions

The interactions between ecological and physical processes are intricate and multifaceted, demonstrating the interconnectedness of biological, chemical, and physical elements in nature. Understanding these interconnections provides valuable insights into ecosystem functionality and highlights how human activities can disrupt these natural processes.

Focus of the Chapter

In this chapter, we delve into the various aspects of decomposition, including the organisms involved, the chemical reactions driving the process, and the physical changes within the soil. We will examine decomposition's role in the global carbon cycle and discuss its significance for soil fertility and health. Additionally, we will explore how environmental changes, whether caused by natural factors or human intervention, influence decomposition processes and their impacts on ecosystems.

7.1. Ecological Perspectives

7.1.1. The Role of Decomposition in the Nutrient Cycle

Nutrient Recycling

Decomposition of dead organic matter is a critical component of the nutrient cycle. The nutrients released, such as nitrogen, phosphorus, and potassium, are returned to the soil, where they become available for plant uptake.

Humus Formation

During decomposition, humus is formed, which stores essential nutrients and enhances soil fertility. Humus acts as a buffer, retaining nutrients and making them accessible to plants over the long term.

Decomposition's role in the nutrient cycle is fundamental to the functioning of ecosystems. By breaking down dead organic matter, such as fallen leaves, decayed plants, and animal carcasses, vital nutrients that would otherwise remain locked in dead biomass are liberated. These nutrients, including nitrogen, phosphorus, and potassium, are indispensable for the growth and development of plants.

Key Nutrients and Their Functions

Nitrogen: Essential for amino acids and proteins, nitrogen plays a crucial role in photosynthesis. Through decomposition, nitrogen is released as ammonium and nitrates, which are easily absorbed by plant roots.

Phosphorus*:* A major component of DNA, RNA, and ATP, phosphorus is vital for energy storage and transfer within cells.

Potassium: Regulates the water balance in plant cells and activates numerous enzymes necessary for plant growth.

Formation and Importance of Humus

Humus formation is another significant outcome of decomposition. Created through microbial and chemical transformation of organic matter, humus represents a stable, long-lasting form of organic carbon. Its presence in soil has numerous positive effects on soil quality and fertility. Humus has an exceptional capacity to retain water and nutrients, releasing them gradually. This capability is particularly crucial in nutrient-poor soils, ensuring a steady supply of nutrients to plants even during periods of scarcity or adverse growth conditions.

Moreover, humus improves soil structure by encouraging the aggregation of soil particles. This enhanced structure promotes better aeration and water infiltration, which, in turn, supports root growth and microbial activity. Increased microbial activity accelerates the breakdown of organic matter and the release of nutrients, creating a positive feedback loop that further boosts soil fertility.

Ecological Impacts of a Healthy Nutrient Cycle

A robust nutrient cycle is vital for ecosystem stability and productivity. Nutrient-sufficient plants are more resilient to diseases and stressors such as drought or pests. These plants, in turn, provide food and habitat for a wide range of animals, from insects to large mammals, that depend on plant biomass. Thus, decomposition supports not only plant life but the entire biological productivity of an ecosystem.

Relevance in Human-Managed Systems

The importance of decomposition in the nutrient cycle extends to human-managed systems like agriculture and forestry. Efforts are often made to optimize natural decomposition processes to enhance soil fertility. For example, adding organic matter such as compost or mulch promotes decomposer activity and supports humus formation.

Conclusion

Decomposition is a complex and dynamic process deeply integrated into ecological and physical systems. Understanding its role in the nutrient cycle provides valuable insights into ecosystem functioning and informs sustainable practices to enhance soil fertility and ecological health. By promoting natural decomposition, ecosystems can maintain their productivity and resilience, supporting a wide array of life forms and ensuring long-term sustainability.

7.1.2. Effects on Soil Biology

Microbial Communities

Decomposition significantly influences the composition and activity of microbial communities in soil. Different microbes are responsible for various stages of decomposition, contributing to humus formation and nutrient availability.

Microbial communities in soil play a pivotal role in the decomposition process and the resulting soil fertility. These communities comprise diverse microorganisms, including bacteria, fungi, actinomycetes, and other microbes, each fulfilling specific roles in breaking down organic matter. The diversity and activity of these microorganisms are critical in determining the speed and efficiency of decomposition, as well as the quality of the humus produced.

At the onset of decomposition, bacteria and fungi dominate, breaking down easily degradable components of organic material, such as sugars, proteins, and lipids. These microorganisms secrete enzymes that dismantle complex molecules into smaller, usable units. For example, certain bacterial species excel in hydrolyzing starch into glucose, while specific fungi efficiently degrade cellulose. These initial decomposition steps lay the foundation for subsequent stages of breakdown.

As decomposition progresses, specialized microbes take over, breaking down more complex and resistant components of organic material. Actinomycetes, a group of filamentous bacteria, are particularly known for their ability to decompose lignin-rich substances. Lignin, a primary constituent of wood and other plant materials, is notoriously difficult to break down. Actinomycetes play a vital role in converting these complex substances into stable humus compounds.

Microbial activity directly impacts soil fertility. Organic compounds are transformed into inorganic nutrients, such as ammonium, nitrates, phosphates, and potassium, which are essential for plant growth. Additionally, microbial metabolic byproducts stabilize soil structure by promoting the aggregation of soil particles and supporting the formation of crumbly textures.

Another critical aspect of microbial communities is their interaction with other soil organisms. Symbiotic relationships, such as those between mycorrhizal fungi and plant roots, significantly enhance nutrient uptake and improve soil structure. Mycorrhizal fungi form networks around plant roots, increasing the surface area for nutrient absorption, which helps plants access more water and nutrients. In return, the fungi receive carbohydrates from the plants, which they use for energy.

The diversity and dynamic interactions of microbes are crucial for soil resilience against environmental stressors. A diverse microbial pool enables soil to better adapt to changes in conditions, such as fluctuations in moisture or temperature.

This resilience is particularly critical in the context of climate change, as soils rich in microbial diversity are more resistant to extreme weather and ensure continuous nutrient cycling.

Summary

Microbial communities are integral to decomposition and play a crucial role in soil fertility and health. Their ability to break down organic matter and release nutrients, coupled with their interactions with other soil organisms, makes them indispensable to the soil ecosystem. Understanding microbial processes and their interconnections helps develop sustainable practices to maintain and enhance soil fertility, ensuring long-term ecosystem productivity.

Soil Fauna

Decomposition supports soil fauna communities, including worms, insects, and other detritivores, which contribute to further decomposition and soil structuring.

Decomposition profoundly impacts soil biology, shaping and sustaining the biodiversity and functionality of soil ecosystems. Microbial communities are particularly significant, acting as the primary agents of decomposition. These communities are highly diverse, comprising bacteria, fungi, actinomycetes, and other microorganisms, each performing specific roles in breaking down organic matter.

At the initial stages of decomposition, bacteria and fungi dominate, efficiently breaking down simple organic compounds such as sugars and amino acids. These early decomposers produce enzymes that penetrate plant cell walls and release the nutrients within. As the decomposition process continues, specialized microbes take over, targeting complex organic molecules like lignin and cellulose. Actinomycetes, a group of filamentous bacteria, excel at breaking down tough plant materials, playing a critical role in final humus formation.

The microbial activity has direct implications for soil fertility and nutrient availability. Through their metabolic processes, microbes convert organic matter into inorganic nutrients that plants can readily absorb. These nutrients are then released into the soil, where they are accessible to plant roots. As a result, microbial communities are vital not only for decomposition but also for maintaining a healthy and productive soil ecosystem.

Decomposition also affects soil fauna, commonly referred to as macrofauna. These include earthworms, insects, mites, and other detritivores that feed on dead organic matter. Earthworms are particularly important, as their burrowing activity aerates the soil and mixes organic material with mineral soil, promoting the formation of fertile topsoil. Their movement also enhances soil porosity, water retention, and root growth.

Insects and other small soil organisms also play a significant role in decomposition by breaking down organic material into smaller pieces, increasing the surface area available for microbial action. This fragmentation accelerates the decomposition process, ensuring that nutrients are rapidly recycled into the soil. The interactions between microbes and macrofauna are thus a cornerstone of the decomposition process, collectively contributing to soil formation and health.

The diversity of soil organisms involved in decomposition results in a complex and stable soil ecology. A well-functioning decomposition system supports various ecological processes, including nutrient cycling regulation, soil structure maintenance, and plant health promotion. These processes are vital not only for natural ecosystems but also for agricultural systems, where soil fertility and productivity depend on soil health.

Conclusion

Decomposition has multifaceted and far-reaching impacts on

soil biology. Understanding these processes allows us to better appreciate soils as living systems and how they are influenced by human activities. Promoting soil biology through practices like adding organic matter and avoiding soil compaction can optimize decomposition processes and improve soil quality over time. Supporting healthy soil biology is, therefore, a key factor in sustainable land use and environmental conservation.

7.1.3. Effects on Plant Growth
Nutrient Supply

Decomposition provides essential nutrients that are critical for the growth and development of plants. A balanced nutrient profile supports healthy plant growth and productivity.

Decomposition plays a pivotal role in the nutrient supply for plants by continuously enriching the soil with key nutrients necessary for their development and vitality. The process breaks down organic material such as dead plant debris, leaves, and animal remains, converting it into inorganic nutrients. These nutrients—nitrogen, phosphorus, and potassium—are fundamental to many physiological processes in plants.

Nitrogen, for instance, is a crucial component of amino acids, proteins, and chlorophyll. Chlorophyll is essential for photosynthesis, the process by which plants convert light energy into chemical energy. A nitrogen deficiency results in stunted growth, yellowing leaves, and reduced plant vitality.

Phosphorus is another indispensable nutrient required for energy transfer within plant cells. It is a key component of adenosine triphosphate (ATP), the molecule that stores and transfers energy in cells. Phosphorus also promotes root development, flowering, and fruit production. A deficiency in phosphorus can delay growth, hinder root development, and reduce flower and fruit yields.

Potassium plays a vital role in regulating water balance within plant cells and activating enzymes involved in various metabolic processes. It enhances photosynthesis, strengthens cell walls, and boosts plant disease resistance. A potassium deficiency may lead to wilting, browning leaf edges, and increased susceptibility to diseases.

Maintaining a balanced supply of these nutrients is critical for healthy plant growth and high productivity. Decomposition ensures that these nutrients are available in forms that plants can easily absorb. By consistently releasing nutrients from decaying organic matter, decomposition increases soil fertility, enabling plants to be well-nourished over extended periods.

In addition to macronutrients like nitrogen, phosphorus, and potassium, decomposition contributes to the availability of micronutrients, which are needed in smaller quantities but are equally essential for plant health. These include elements such as iron, manganese, zinc, copper, and boron. These micronutrients play crucial roles in numerous biochemical reactions and contribute to the vitality of plants.

Another significant outcome of decomposition is the creation of a favorable microbial environment in the soil. Microorganisms involved in decomposition not only break down organic materials but also foster symbiotic relationships with plants. One notable example is mycorrhizae, a symbiosis between fungi and plant roots that enhances nutrient uptake. Mycorrhizal fungi extend the root network of plants, helping them absorb water and nutrients more efficiently, especially in nutrient-poor soils.

Overall, decomposition illustrates the interconnectedness of nutrient cycling and plant growth. By continuously releasing and providing nutrients, decomposition significantly contributes to soil fertility. A deep understanding of these processes enables the optimization of agricultural and horticultural practices to improve soil quality and increase plant productivity. Promoting decomposition through organic fertilization, composting, and other sustainable practices is instrumental in ensuring the long-term health and productivity of ecosystems.

Soil Structure

Humus formation improves soil structure by increasing water retention capacity and enhancing soil aeration. These improvements create more favorable conditions for roots and plant growth.

Decomposition is vital for plant growth as it enriches the soil with key nutrients essential for plant development and health. The breakdown of organic material by microorganisms leads to the release of nutrients such as nitrogen, phosphorus, and potassium, which are indispensable for processes like photosynthesis, protein synthesis, and other vital functions.

Nitrogen, for example, is a primary component of chlorophyll, the molecule responsible for photosynthesis. Without sufficient nitrogen, plants cannot produce adequate chlorophyll, resulting in stunted growth and diminished energy production. Phosphorus is equally critical, facilitating energy transfer within plant cells. It is a fundamental part of ATP (adenosine triphosphate), the principal energy molecule in cells. Potassium, on the other hand, is crucial for maintaining water balance in plant cells and activating enzymes essential for plant growth and development.

The continuous supply of these nutrients through decomposition ensures optimal growing conditions for plants. A well-balanced soil nutrient profile not only promotes robust plant growth but also enhances their resilience against diseases and environmental stresses such as drought and pests. Plants that receive adequate nutrients typically exhibit better root development, stronger foliage, and higher productivity.

In addition to nutrient supply, decomposition has a significant impact on soil structure, which in turn positively affects plant growth. The formation of humus, a byproduct of decomposition, plays a crucial role in improving soil structure. Humus increases the soil's water retention capacity, which is particularly valuable in arid regions. Soils rich in humus can store more water and release it gradually to plant roots, reducing drought stress.

Moreover, humus promotes the aggregation of soil particles, leading to better soil aeration. A well-aerated soil structure allows roots to penetrate deeper, enabling more efficient absorption of water and nutrients. Enhanced root systems increase plant stability and resilience against environmental challenges. Soil aeration also benefits microbial communities involved in decomposition, as many of these microorganisms require oxygen for their metabolic processes.

The physical changes in the soil resulting from decomposition and humus formation create an environment that is highly conducive to plant growth. These benefits extend beyond natural vegetation to agricultural crops. Farmers and gardeners can enhance soil quality and boost plant yields by encouraging decomposition and enriching soil with organic material.

An additional advantage of improved soil structure is the reduction of erosion. Humus-rich soils are less susceptible to erosion by wind and water because the soil particles are more cohesive. This helps preserve the nutrient-rich topsoil, essential for plant growth, and prevents the loss of valuable organic material.

In summary, decomposition is critical for plant growth through its role in nutrient supply and soil structure improvement. The release of essential nutrients and the physical transformation of soil create an environment in which plants can thrive. A deeper understanding of these processes allows for targeted measures to enhance plant growth and support sustainable agricultural practices.

7.1.4. Ecological Dynamics
Biodiversity

Decomposition enhances biodiversity by creating habitats for various microbes and soil organisms. This diversity supports ecological stability and resilience in ecosystems.

Decomposition is pivotal in promoting biodiversity within soils and ecosystems at large. As organic material breaks down, numerous microhabitats and ecological niches are formed, providing ideal conditions for a wide range of organisms. Microbes, including bacteria, fungi, and actinomycetes, thrive in decomposing material, each specializing in different substrates and decomposition phases. This specialization fosters high microbial diversity.

These microbial communities interact intricately with larger soil organisms, such as worms, mites, insects, and other detritivores. Each organism plays a unique role in the decomposition process and contributes to nutrient cycling. For example, earthworms fragment organic material, increasing its surface area for microbial activity. Insects and other arthropods further break down the material and distribute it throughout the soil, ensuring uniform accessibility for microbes.

The biodiversity fostered by decomposition contributes to ecosystem stability and resilience. Diverse ecosystems are better equipped to respond to disturbances, whether natural events like floods or droughts or human activities such as agriculture and urbanization. The wide range of organisms in these ecosystems ensures that essential ecological functions are maintained even if specific species or groups are impacted. This stability is critical for the long-term health and functionality of ecosystems.

Moreover, the microhabitats created by decomposition provide refuge for rare and specialized species. These species contribute to genetic diversity and often fulfill unique ecological roles vital to ecosystem balance. Thus, by promoting biodiversity, decomposition not only stabilizes existing ecosystems but also aids in preserving overall biological diversity.

Carbon Storage

Decomposition impacts the Earth's carbon balance. While some carbon is released into the atmosphere as carbon dioxide, a significant portion is stored in the soil as organic matter, contributing to carbon sequestration and climate mitigation.

Decomposition is a key player in the global carbon cycle, influencing both carbon release and storage. When organic material breaks down, carbon is released as carbon dioxide (CO_2), increasing atmospheric carbon levels. However, a substantial fraction of decomposed material is retained in the soil as stable organic matter, particularly as humus.

Humus, a complex mixture of partially decomposed plant and animal material, is relatively stable and can persist in the soil for extended periods. By storing carbon in the form of humus, soils act as carbon sinks, significantly reducing the concentration of CO_2 in the atmosphere. This is especially critical in the context of climate change, as soil carbon sequestration provides a natural method for mitigating greenhouse gas emissions.

The storage of carbon in the soil also brings additional ecological benefits. Soils rich in organic matter exhibit improved structure and increased water retention capacity, creating favorable conditions for plant growth and enhancing soil fertility. Enhanced soil structure also helps prevent erosion and improves water infiltration, further strengthening ecosystems against extreme weather events.

The long-term sequestration of carbon in soils is therefore a vital component of climate mitigation strategies and sustainable land management. Encouraging decomposition and increasing soil humus content can contribute to carbon sequestration, positively affecting the global carbon balance and climate.

Summary

Decomposition plays a fundamental role in ecological dynamics by promoting biodiversity and contributing to carbon storage. These processes bolster the stability and resilience of ecosystems and are integral to global climate protection. By fostering biodiversity through habitat creation and supporting carbon sequestration in soils, decomposition links biological activity with broader environmental processes.

Understanding decomposition and its ecological impacts enables the development of sustainable practices that safeguard biodiversity and combat climate change. This knowledge is crucial for maintaining ecosystem functionality and achieving a balance between ecological health and human activities.

7.2. Physical Perspectives

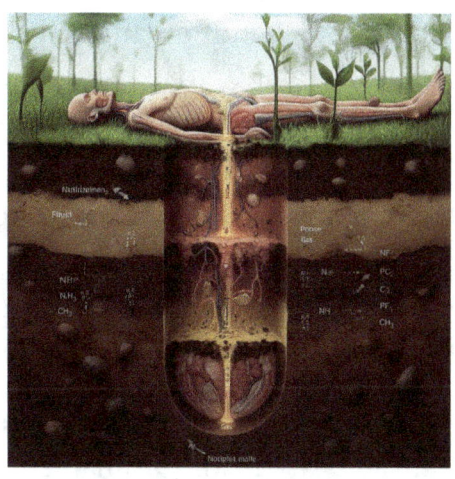

7.2.1. Temperature and Moisture Conditions

Impact of Temperature

The rate of decomposition is highly dependent on temperature. High temperatures accelerate microbial activity and enzymatic processes, while low temperatures slow down decomposition. This directly influences the speed of humus formation and nutrient availability.

Temperature is a critical factor in the decomposition of organic materials within soil. At higher temperatures, microbial activity intensifies as microorganisms thrive, grow, and reproduce more rapidly in warmer environments. Enzymes produced by these microorganisms operate more efficiently under elevated temperatures, leading to the accelerated breakdown of complex organic compounds. This results in faster nutrient release and quicker humus formation. For instance, tropical regions, with consistently high temperatures, exhibit significantly faster decomposition rates compared to colder climates.

Conversely, low temperatures suppress microbial activity and enzymatic processes, which are vital for decomposition. In colder regions or during winter months in temperate zones, the breakdown of organic substances slows dramatically. This causes organic materials, such as leaves and wood, to accumulate over years and convert to humus at a much slower pace.

Temperature dependency also directly impacts nutrient availability in soil. In warmer conditions or seasons, plants can access nutrients more quickly due to the faster pace of decomposition, supporting robust growth provided other essential factors like water and light are sufficient. In colder periods, reduced nutrient release limits plant growth as fewer nutrients are readily available.

Seasonal fluctuations in temperature further affect microbial activity and decomposition dynamics. In temperate climates with significant seasonal variation, decomposition processes peak during summer when microbial activity is at its highest. These seasonal differences influence the annual nutrient cycle, plant growth, and soil fertility.

Moisture Conditions

Moisture significantly affects decomposition processes. High moisture levels promote microbial activity and the breakdown of organic materials, while dry conditions slow decomposition and lead to the accumulation of less-decomposed material.

Soil moisture is another critical determinant of decomposition rates. Water is essential for most biological and chemical processes, including the functions of microorganisms involved in decomposition. In moist soils, microbes can efficiently absorb nutrients and carry out metabolic activities, facilitating the rapid breakdown of organic material.

Regions with high moisture levels, such as tropical rainforests or temperate forests during rainy seasons, experience accelerated decomposition. Organic materials, including fallen leaves, dead plants, and animal remains, decompose swiftly, and the released nutrients are quickly reabsorbed by plants.

This results in fertile soils that support dense and diverse vegetation.

In contrast, decomposition slows considerably in arid and semi-arid regions where water is scarce. Microbial activities and enzymatic processes are significantly hindered under dry conditions. Organic materials remain largely intact for extended periods, with nutrient release occurring at a much slower rate. This leads to the accumulation of partially decomposed materials, which may only break down further when adequate moisture becomes available. The slow decomposition in dry environments affects soil fertility and often restricts plant growth.

Moisture also indirectly impacts soil temperature and decomposition efficiency. Moist soils tend to maintain more stable temperatures because water acts as a heat buffer, storing and transferring warmth. This stability fosters consistent microbial growth and activity. Dry soils, however, are more prone to temperature fluctuations, which can further hinder microbial processes.

Conclusion

Both temperature and moisture are pivotal physical factors shaping the decomposition of organic materials in soil. Their interplay determines the efficiency and speed of decomposition processes, humus formation, and nutrient availability for plants. A comprehensive understanding of these factors and their interactions is essential for sustainable soil management and enhancing soil fertility across various climates.

7.2.2. Physical Weathering Processes

Mechanical Weathering

Physical processes such as frost action and thermal weathering significantly impact the decomposition of organic material by breaking it into smaller fragments that microbes can more easily digest.

Mechanical weathering is a crucial factor in the decomposition process, as it physically fragments organic material, thereby increasing the surface area available for microbial activity. A prime example is frost action, where water seeps into pores within organic material, freezes, and expands. This expansion generates internal stresses that eventually fracture the material into smaller pieces. These fragments expose more surface area for microbial colonization, accelerating the decomposition process.

Thermal weathering, another critical mechanism, occurs when temperature fluctuations cause organic material to expand and contract repeatedly. This cyclical stress can lead to cracks and fissures that further break down the material. In arid and mountainous regions, where day-to-night temperature variations are extreme, thermal weathering is particularly pronounced. The fragmentation resulting from this process enhances the accessibility of organic material to microbes, thereby supporting faster microbial degradation.

Mechanical weathering is not limited to visible fragmentation. On a microscopic level, mechanical forces also play a role. For instance, plant roots infiltrate tiny cracks and crevices, exerting pressure that physically pries the material apart. Similarly, soil organisms like earthworms and insects contribute to mechanical breakdown through their movement and feeding activities. These organisms grind and distribute organic matter, making it more accessible for microbial communities.

Fragmentation not only hastens decomposition but also facilitates the release of trapped nutrients. Increased surface area supports more intensive microbial activity, promoting faster transformation of the material into humus-forming substances. This efficient nutrient release is critical for supporting plant growth and maintaining soil fertility.

Chemical Weathering

Chemical weathering processes, such as the interaction of organic materials with oxygen and water, further drive decomposition and mineralization. These reactions release nutrients and influence soil structure.

Chemical weathering complements mechanical processes by breaking down organic materials at the molecular level. One of the primary mechanisms of chemical weathering is oxidation, where organic compounds react with oxygen to form new substances that are more easily degraded. This reaction often releases essential nutrients such as carbon, nitrogen, and sulfur in forms that plants can readily absorb.

Water is a key player in many chemical weathering processes. Hydrolysis, in which water breaks chemical bonds in organic molecules, is a common pathway for decomposition. Water also acts as a solvent, dissolving nutrients and facilitating their transport through the soil. In moist environments, these chemical reactions are particularly active, increasing nutrient availability and promoting more rapid decomposition.

Acid weathering is another example, where organic acids produced by plant roots and microorganisms attack both organic and mineral components of the soil. These acids dissolve chemical bonds, releasing nutrients that plants can use. Acid weathering is especially significant in acidic soils, where it enhances nutrient availability and improves soil structure.

Chemical weathering also contributes to mineralization, the transformation of organic substances into inorganic minerals. This process is a critical step in nutrient cycling, as plants primarily absorb nutrients in inorganic forms. For instance, phosphorus and potassium are mineralized into plant-available forms, supporting growth and development. The speed and efficiency of mineralization depend on the soil's chemical properties and the composition of the organic material.

In addition to nutrient release, chemical weathering influences soil structure. By transforming organic material into stable humus compounds, it promotes the aggregation of soil particles. This aggregation enhances soil structure, increases water retention, and reduces erosion susceptibility. Well-structured soil provides better conditions for plant growth by efficiently storing and transporting water and nutrients.

Conclusion

Mechanical and chemical weathering processes are integral to decomposition, often working synergistically to break down organic material into forms that microbes can more readily digest. Together, these processes ensure efficient nutrient release and transformation, which are vital for plant growth and soil fertility. Understanding these mechanisms provides valuable insights for developing sustainable soil management practices that enhance long-term soil quality and ecological health.

7.2.3. Energy Transfer and Loss

Energy Transformation

During decomposition, chemical and stored energy are released in the form of heat and gases. This energy is utilized by microbes and soil organisms, influencing soil temperature and altering soil chemistry.

Decomposition of organic material is not solely a biological process but also an energetic one. As organic matter, such as plant residues or animal remains, breaks down, the chemical bonds within it are disrupted, releasing stored energy. This energy is emitted as heat and gases, including carbon dioxide (CO_2), methane (CH_4), and ammonia (NH_3).

The chemical energy stored in organic substances is converted through microbial activity, enzymatic reactions, and other chemical processes. Microorganisms involved in decomposition harness this energy to sustain their metabolic functions, including growth, reproduction, and cellular maintenance. As these microbes break down complex compounds into simpler forms, energy is released, some of which dissipates as heat into the surrounding soil.

This released heat can significantly impact the soil's thermal balance, particularly in contained or insulated systems like compost piles or waste heaps. In these environments, decomposition can cause temperatures to rise substantially. For example, within a compost pile, microbial activity may elevate the temperature to levels that accelerate the breakdown of organic material. Higher temperatures also encourage the proliferation of thermophilic (heat-loving) microbes, which further intensify the decomposition process. This creates a feedback loop where heat stimulates microbial activity, leading to faster decomposition and additional heat production.

The energy released during decomposition is also evident in the production of gases such as CO_2 and CH_4. Carbon dioxide is primarily produced during aerobic decomposition, where microbes degrade organic material in the presence of oxygen. Conversely, methane is generated under anaerobic conditions, such as in compacted soils or waterlogged environments. The type and quantity of gases emitted depend on specific decomposition conditions, including oxygen availability and the nature of the organic material being decomposed.

The release of these gases has broader environmental implications. Carbon dioxide contributes to the atmospheric greenhouse effect, while methane, a more potent greenhouse gas, has a far stronger warming potential. Proper management of decomposition processes, particularly in large-scale organic waste systems, is essential to minimize these emissions and their environmental impacts. This includes controlling conditions to favor aerobic decomposition and limiting the production of methane in anaerobic environments.

Heat Generation

The heat generated during decomposition can influence soil temperature, particularly in insulated or contained systems such as compost heaps or waste piles. This phenomenon can significantly impact microbial activity, nutrient availability, and overall decomposition efficiency.

The heat produced by microbial activity during decomposition has substantial effects on the soil's temperature profile. In enclosed systems like compost heaps, biogas plants, or waste piles, microbial metabolism generates considerable amounts of heat. This heat is a byproduct of the energy transformations occurring as microbes break down organic material.

In composting systems, for instance, the temperature within the pile can rise to 40–70°C (104–158°F) due to the exothermic reactions of decomposition. This elevated temperature has several beneficial effects. It promotes the activity of thermophilic microbes, which specialize in breaking down more resistant organic compounds such as cellulose and lignin. These microbes not only accelerate decomposition but also generate additional heat, creating an optimal environment for further breakdown of organic material.

The heat in compost piles also serves a sanitizing function by destroying pathogens and weed seeds, improving the quality of the resulting compost. However, if not managed properly, excessive heat can inhibit microbial diversity and slow the decomposition process. To ensure optimal conditions, compost systems are regularly aerated and monitored for temperature fluctuations.

In soil, heat released by decomposition can indirectly affect other physical and chemical properties. For example, elevated temperatures may accelerate chemical reactions, such as nutrient mineralization, making nutrients more readily available to plants. However, increased heat can also lead to moisture loss through evaporation, potentially reducing soil water availability. This dual effect highlights the importance of balancing decomposition processes to maintain soil health.

Waste piles and landfills also experience significant heat generation. In these environments, temperature increases can impact the structural integrity of the waste material and influence gas exchange dynamics. Effective waste management strategies often involve temperature control measures, such as ventilation or periodic turning of the waste, to regulate microbial activity and prevent excessive methane production.

Conclusion

Energy transfer and loss during decomposition are dynamic processes involving the release of heat and gases. These processes influence decomposition efficiency, soil chemistry, and environmental conditions. The heat generated fosters microbial activity and affects nutrient cycling, while gases like CO_2 and CH_4 have critical implications for climate change. Understanding these interactions is essential for optimizing decomposition processes, managing organic waste sustainably, and mitigating environmental impacts. By leveraging this knowledge, we can better harness decomposition for ecological and agricultural benefits while minimizing its contribution to greenhouse gas emissions.

7.2.4. Effects on Soil Physics

Soil Structure

Decomposition has a profound impact on the physical structure of soil. The breakdown of organic material leads to the formation of humus, which enhances soil structure and promotes the aggregation of soil particles.

The physical structure of soil is a critical determinant of its fertility and overall functionality within an ecosystem. As organic materials such as plant debris, leaves, and other organic residues undergo decomposition, they are broken down into smaller particles through the activity of microbes, fungi, and soil organisms. These processes fundamentally alter the composition and structure of the soil.

Humus formation, a key outcome of decomposition, plays an essential role in improving soil structure. This stable organic substance acts as a binding agent, bringing soil particles together to form aggregates. These aggregates contribute to a loose, crumbly soil structure that facilitates aeration and supports root growth. The creation of such aggregates not only enhances soil stability but also improves its resilience against physical erosion.

Aggregated soil is less prone to erosion by wind and water, as the tightly bound particles resist being easily displaced. This preservation of the topsoil—the most nutrient-rich layer—is vital for maintaining soil fertility. Additionally, the enhanced soil structure improves porosity, allowing for better air circulation and water infiltration. Increased porosity creates an optimal environment for root systems and soil organisms to thrive. This makes the soil easier to work with in agricultural contexts, contributing to efficient farming practices.

Water Retention

The organic matter and humus content significantly influence the soil's water-holding capacity. A higher concentration of humus improves water retention and reduces surface runoff.

The soil's ability to retain water is a critical factor for plant hydration and overall soil quality. Humus, the stable end product of organic matter decomposition, has exceptional water-holding properties due to its high specific surface area and capacity for water absorption. These characteristics enable the soil to retain larger amounts of water while minimizing loss through evaporation or surface runoff.

Soils rich in humus can store and hold water more effectively, providing plants with a reliable supply, especially during dry periods or drought conditions. Enhanced water retention reduces the frequency and need for irrigation, leading to cost savings in agricultural operations. Improved water availability also strengthens plants' resistance to drought stress and boosts overall crop productivity.

Moreover, better water retention reduces surface runoff. When soil can absorb and store more water, less excess water flows across the surface, lowering the risk of erosion and soil loss. This not only conserves the topsoil but also ensures that more water remains available for root uptake, further enhancing plant hydration.

Improved water retention is closely tied to the soil's porosity, which is enhanced by humus formation. Humus contributes to aggregate formation, increasing the number and size of pores in the soil. These pores allow water to infiltrate and be stored within the soil matrix. Enhanced porosity benefits both plants and soil stability by reducing waterlogging and improving drainage while ensuring that adequate moisture is retained.

Summary

In summary, the decomposition of organic material has extensive impacts on the physical properties of soil. The formation of humus enhances soil structure by promoting aggregation and increasing porosity. Simultaneously, a higher concentration of humus improves the soil's water-holding capacity, leading to better water retention and reduced surface runoff. These physical changes result in improved soil quality, increased plant productivity, and greater resilience against erosion and drought stress. Understanding these processes is essential for effective soil management and the sustainable use of soil resources.

Energy and Transience -The Journey After Death

7.3. Practical Applications and Research

7.3.1. Agricultural Applications

Composting

The deliberate application of decomposition processes in composting enhances soil fertility and nutrient availability. Composting utilizes controlled decomposition to produce valuable organic matter that can be used in agriculture.

Composting is a practical example of harnessing natural decomposition processes to improve soil quality. By combining organic materials such as kitchen scraps, garden waste, and agricultural residues in a controlled environment, decomposition is managed under optimized conditions of temperature, moisture, and aeration. This ensures the efficient activity of microbes and accelerates the breakdown of materials.

Through composting, complex organic compounds are converted into simpler, more stable forms known as humus. This humus is nutrient-rich and exhibits excellent properties for enhancing soil structure. It promotes soil aggregation, improving porosity and water retention, which supports healthy root growth and boosts plant productivity.

Additionally, composting significantly reduces the volume of organic waste, turning it into a beneficial product instead of landfill material. This practice mitigates environmental stress caused by waste disposal while also releasing energy during the composting process, further fueling microbial activity that continues to improve nutrient release and soil enhancement.

The quality of the compost is highly dependent on the materials used and the conditions maintained during the composting process. Careful selection of inputs, regular turning of the compost pile, and proper moisture control are essential for producing high-quality compost that is nutrient-rich and effective in improving soil properties.

Composting also plays a vital role in organic farming. By utilizing self-produced compost, farmers can reduce reliance on chemical fertilizers and adopt more sustainable agricultural practices. This not only enhances soil quality but also supports ecological balance and the long-term health of agricultural systems.

Soil Management

An in-depth understanding of decomposition and its impact on soil enables the development of sustainable soil management practices that preserve soil fertility and health over the long term.

Knowledge of decomposition processes is critical for effective soil management as it directly informs the design of sustainable farming strategies. Sustainable soil management practices leverage natural decomposition to maintain and enhance soil fertility. These include the targeted use of organic fertilizers, mulching, and cover cropping.

Organic fertilizers such as compost or manure improve soil fertility by continuously releasing nutrients and fostering humus formation. These nutrients are made available to plants, ensuring the soil remains healthy and productive. Mulching, which involves applying a layer of organic material on the soil surface, protects the soil from erosion, reduces evaporation, and supports humus development.

Cover cropping, or growing cover crops between main crop cycles or as a precursor to planting, is another sustainable practice that enriches soil organic matter, enhances structure, and promotes decomposition. Cover crops can also help sequester nutrients in the soil and stabilize its structure, reducing erosion and increasing water-holding capacity.

A comprehensive understanding of decomposition processes allows farmers to implement measures that prevent soil erosion and nutrient loss. By selecting and applying appropriate management practices, the soil can be stabilized, and nutrient supply optimized, leading to increased productivity and sustainability in agriculture.

Research in this area continuously explores new approaches and technologies to improve soil management. This includes developing innovative composting techniques, utilizing new types of organic fertilizers, and studying the impact of different farming practices on soil quality. Such research provides evidence-based recommendations for farmers and supports the development of sustainable solutions to soil management challenges.

Summary

In conclusion, applying the principles of decomposition in agriculture is essential for enhancing soil fertility and sustainability. Through targeted composting and thoughtful soil management, farmers can maintain soil quality, boost productivity, and adopt environmentally friendly farming practices. Continuous understanding and further research in this field are vital for identifying best practices and ensuring the long-term sustainable management of soil resources.

7.3.2. Environmental Research
Carbon Balances

The study of decomposition processes is critical for assessing carbon balances and understanding the role of soil in global carbon storage. Such insights are vital for analyzing climate impacts and developing targeted climate protection strategies.

Decomposition is a cornerstone of the carbon cycle, shaping how carbon is released as CO_2 or stored as stable humus. By analyzing these processes, scientists can quantify how much carbon is sequestered in soil versus how much is emitted back into the atmosphere through microbial activities. This understanding is fundamental for modeling regional and global carbon balances with precision.

Soil carbon content serves as a key indicator of soil quality and its potential to act as a carbon sink. Through decomposition, organic matter is broken down, with some carbon being emitted as CO_2 and another portion remaining as stable humic substances. These stable forms of carbon play a pivotal role in reducing atmospheric CO_2 levels, thereby influencing climate dynamics and serving as a critical buffer against global warming.

Research also delves into how agricultural and land management practices influence soil carbon storage. For example, reduced tillage, the use of cover crops, and organic amendments such as compost have been shown to enhance carbon sequestration in soils. Such practices not only enrich soil fertility but also contribute to climate mitigation by capturing atmospheric carbon.

Additionally, studies examine how soil types, climatic conditions, and land use impact the decomposition process and associated carbon dynamics. These findings inform the development of more accurate climate models and policy recommendations aimed at reducing greenhouse gas emissions. Insights gained are integral to implementing carbon capture measures highlighted in international agreements like the Paris Climate Accord.

Soil Restoration

Decomposition research also informs efforts to restore and rehabilitate soils degraded by erosion, pollution, or other environmental challenges, providing a foundation for reconstructing fertile and functional soils.

Soil degradation, caused by processes like erosion, nutrient depletion, and contamination, poses a significant threat to ecosystem health and agricultural productivity. Decomposition plays a pivotal role in reversing such damage by driving the regeneration of organic matter and supporting humus formation, which are essential for restoring soil fertility.

Restoration strategies often include the introduction of organic materials to accelerate decomposition and boost humus levels. For example, applying organic compost to eroded areas can rebuild topsoil layers and improve water retention. In heavily polluted soils, specialized amendments, such as bioengineered compost, are employed to neutralize toxins and reinvigorate microbial activity, thus aiding in the breakdown of harmful substances.

Research in this field explores advanced methods for restoring soil functionality and enhancing ecological resilience. Techniques such as planting cover crops, which promote decomposition and contribute to humus accumulation, are being refined to optimize their effectiveness. These crops stabilize soil structures, reduce erosion, and improve water infiltration, creating conditions conducive to long-term soil health.

Innovative technologies, such as the application of biochar and the introduction of microbial inoculants, are also under investigation. These methods aim to accelerate nutrient cycling, enhance soil microbial diversity, and improve the capacity for carbon storage. Such approaches offer promising solutions for large-scale soil restoration efforts while minimizing environmental impacts.

Soil restoration is closely tied to biodiversity conservation and habitat recovery. By improving soil quality and restoring its natural functions, ecosystems can regain their balance, supporting the diverse plant and animal species that depend on healthy soil. This not only strengthens ecological integrity but also ensures the sustainable use of land resources.

Summary

Decomposition processes are at the heart of environmental research, offering critical insights into carbon balances and soil restoration. Understanding how soils sequester carbon and how degradation can be reversed informs the development of effective climate mitigation strategies and soil rehabilitation techniques. Ongoing research in this field contributes to sustainable land management practices, ensuring that soil resources are preserved and utilized efficiently for future generations.

7.4. Summary

Decomposition is a fundamental process with profound impacts on ecological and physical systems. Its influence spans various environmental dimensions, including nutrient cycles, soil biology, plant growth, and physical factors such as temperature, moisture conditions, weathering, and energy transfer. Understanding decomposition comprehensively is crucial for sustainable soil management, the improvement of agricultural practices, and the study of environmental changes.

Ecological Perspectives of Decomposition

From an ecological standpoint, decomposition is vital for nutrient cycles. It ensures that nutrients like nitrogen, phosphorus, and potassium are returned to the soil from dead organic material, where they can be absorbed by plants. This process promotes humus formation, which enhances soil fertility by acting as a buffer that stores nutrients and makes them available to plants over the long term. This nutrient recycling creates a healthy, productive ecosystem by laying the foundation for continued plant growth and development.

Decomposition also has a significant impact on soil biology. Microbial communities in the soil, responsible for various stages of decomposition, play a pivotal role in humus formation and improving nutrient availability. Additionally, decomposition supports the soil fauna, such as worms and insects, which contribute to further decomposition and soil structuring. This biological activity enhances soil health and the efficiency of nutrient cycling.

Plant growth benefits directly from decomposition. The nutrients released are essential for plant development and productivity. A balanced supply of these nutrients ensures robust plant growth and high yields. Furthermore, humus formation improves soil structure, increases water retention capacity, and enhances soil aeration, creating ideal conditions for root development and further supporting plant growth.

Physical Perspectives of Decomposition

On a physical level, decomposition influences soil temperature and moisture conditions. The speed of decomposition is highly temperature-dependent, with higher temperatures accelerating microbial activity and enzymatic processes, while lower temperatures slow them down. Moisture also plays a critical role: high moisture levels promote decomposition, while dry conditions hinder it. These physical factors directly affect the rate of humus formation and nutrient availability.

Decomposition also affects physical weathering processes. Mechanical weathering, such as frost action, breaks organic material into smaller fragments, making it easier for microbes to decompose. Chemical weathering, including reactions with oxygen and water, further contributes to decomposition and mineralization, enhancing nutrient release and improving soil structure.

Energy transfer during decomposition, particularly the release of heat, impacts soil temperature dynamics. In enclosed systems like compost heaps, this heat can significantly alter the temperature range, influencing microbial activity and the decomposition rate.

Practical Applications and Research

Insights from decomposition research have practical applications in agriculture and environmental studies. In agriculture, composting enhances soil fertility and nutrient supply, while an in-depth understanding of decomposition informs sustainable soil management practices. These practices help maintain long-term soil fertility, prevent erosion, and reduce nutrient loss.

In environmental research, decomposition plays a critical role in assessing carbon balances and global carbon storage. Such studies are vital for understanding climate impacts and developing effective climate protection strategies. Additionally, research on soil restoration leverages decomposition processes to rehabilitate soils degraded by erosion or pollution, improving soil quality and structure.

Conclusion

In conclusion, decomposition is a multifaceted process with extensive ecological and physical implications. It is central to nutrient recycling, soil biology, plant growth, and the physical dynamics of the environment. Practical applications of decomposition research enhance soil management, promote sustainable agriculture, and contribute to environmental conservation. A deep understanding of decomposition processes is vital for developing strategies to preserve and improve environmental resources for future generations.

Energy and Transience - The Journey After Death

Chapter 8:
Philosophical and cultural considerations

Death and decomposition are not solely biological processes; they are deeply embedded in philosophical and cultural considerations. The ways in which different cultures and philosophies perceive death and decomposition significantly shape humanity's attitudes toward life and mortality. This chapter delves into the philosophical reflections on death and decomposition and explores cultural practices and rituals that address these profound themes.

Death, as inevitable as it is, remains a subject of great complexity and profound contemplation. Across cultures, death is often seen not just as the end of life but as a transition into another state of existence—a moment imbued with spiritual and existential significance. Philosophies, religions, and scientific perspectives have long sought to explain death and what follows it. These perspectives provide diverse insights into the human experience and the meanings we assign to life and its cessation.

Philosophical Perspectives on Death

In philosophy, death is frequently regarded as the ultimate enigma of life. Existentialist thinkers like Jean-Paul Sartre and Martin Heidegger have offered profound insights into the nature and significance of death. For Sartre, death was a definitive end—a conclusion without continuation—that gives life its urgency and meaning. Heidegger, on the other hand, considered death an integral aspect of existence, shaping the essence of human being. He argued that the awareness of one's mortality leads to a more authentic way of living by compelling individuals to confront the finitude of their existence.

For many philosophical traditions, death is a lens through which the purpose and value of life are examined. Stoicism, for instance, views death as a natural and inevitable event that should be accepted without fear. By contemplating death regularly, the Stoics believed, one could cultivate resilience, clarity, and a focus on living virtuously. In contrast, metaphysical philosophies, like those found in certain strands of Hinduism and Buddhism, frame death as a stage in a larger cycle of rebirth and spiritual evolution. These perspectives often emphasize the impermanence of life and the importance of detachment from worldly attachments.

Cultural Rituals and Practices

Culture plays a pivotal role in shaping our understanding of death and decomposition. Each society develops its own rituals and traditions to cope with death and integrate it into its social and spiritual frameworks. In many indigenous cultures, death is seen as a continuation of life in another form. Rituals honoring ancestors and maintaining a connection between the living and the dead are widespread. These practices highlight an understanding of death as part of a broader cycle of life, death, and rebirth.

In contrast, the modern Western approach to death has evolved over time. Historically, death was an intimate part of daily life, often occurring at home and within the presence of family. However, in contemporary Western societies, death has increasingly been relegated to hospitals and care facilities, distancing it from everyday awareness. Funeral rites and mourning practices have similarly shifted toward more individualized and, at times, commercialized expressions of farewell.

Religious traditions offer additional layers of meaning and ritual around death. In Christianity, for example, death is often viewed as a passage to eternal life, with funeral ceremonies emphasizing the hope of resurrection. In Islam, the rituals surrounding death underscore the return to God, with practices like swift burial and prayers reflecting humility and submission to divine will. In Japan, Buddhist traditions incorporate elaborate rites to ensure the safe passage of the deceased's spirit, demonstrating a blend of religious devotion and cultural heritage.

Modern Reflections on Mortality

Modernity has brought new dimensions to how societies engage with death. Advances in medical science, for instance, have extended life expectancy and shifted the experience of death from an inevitable part of family life to a managed event in institutional settings. This transition has prompted philosophical and ethical debates about the nature of dying, palliative care, and the right to a dignified death. Movements advocating for euthanasia and assisted dying have further complicated these discussions, reflecting evolving attitudes toward autonomy and the sanctity of life.

At the same time, there has been a resurgence of interest in natural and ecological approaches to death. Practices like green burials and human composting challenge conventional funeral practices, aligning the treatment of the body after death with sustainability and environmental consciousness. These approaches reconnect the idea of decomposition with the broader ecological cycles of nature, framing death as a return to the earth rather than a disruption of life's continuity.

The Interplay of Philosophy and Culture

Philosophical and cultural approaches to death are deeply intertwined, reflecting and shaping each other. Philosophical theories about mortality often inform cultural attitudes and rituals, while cultural practices provide a lived context for philosophical reflections. For instance, existentialist ideas about embracing mortality resonate in the traditions of cultures that honor ancestors and celebrate the cyclical nature of life and death.

Conversely, cultures that treat death as a taboo or isolate it from daily life may foster existential anxieties, leading to philosophical inquiries about the fear of death and the search for meaning in an impermanent world. These interconnections demonstrate how philosophical and cultural perspectives together enrich our understanding of mortality and decomposition.

Conclusion

This chapter explores how our understanding of death and decomposition extends beyond biological realities, shaped by deeply rooted cultural and philosophical beliefs. By examining the diverse ways humans have grappled with mortality throughout history, we gain insight into the universal and varied responses to the inevitability of death. Through this lens, we can better appreciate how these perspectives influence not only our individual lives but also the collective practices and values of societies worldwide.

8.1. Philosophical reflections on Death

8.1.1. Philosophical Views on Death

Existential Perspectives

Existentialist philosophers like Jean-Paul Sartre and Albert Camus regard death as a central element of human existence. They see death as the ultimate manifestation of life's absurdity and view the confrontation with mortality as a key to authenticity and finding meaning.

Existentialists emphasize that death is an inescapable reality, one that underscores the urgency and intensity of life itself. Sartre, for example, argues that death represents the ultimate limitation of human freedom, as it marks the definitive end of our existence. This finitude compels us to live consciously and authentically, recognizing that we have only a finite amount of time to realize our potential.

Albert Camus, another prominent existentialist, interprets death as the ultimate evidence of life's absurdity. In his famous essay *The Myth of Sisyphus*, he likens human existence to the eternal, futile labor of Sisyphus, who is condemned to roll a boulder up a hill only for it to roll back down. Yet, Camus finds a sense of defiance and meaning in the conscious decision to persist despite life's absurdity. This rebellion against the meaningless can transform life into a profound and personal act of creation.

By grappling with death and accepting its inevitability, existentialists argue, we can lead an authentic life. Acknowledging the finiteness of our existence allows us to reassess our choices and actions, enabling us to live a more fulfilling and meaningful life.

Stoic Philosophy

Stoics like Epictetus and Seneca advocate for accepting death as part of the natural order. They regard the ability to embrace death with equanimity as a path to inner peace and wisdom. Death, in their view, is an inevitable part of the cosmic process that should be met with acceptance.

Stoic philosophy teaches that we can only control what is within our power, and death lies beyond this domain. Epictetus highlights that death is a natural and necessary aspect of life, and wisdom lies in meeting it without fear or resistance. Accepting the inevitability of death leads to deep inner tranquility, as we align ourselves with the natural order rather than opposing it.

Seneca, another influential Stoic, asserts that death is not inherently harmful; it merely marks the cessation of our physical existence. He advises treating death as a natural event and dismissing the fear it often provokes. For Stoics, death serves as a reminder of life's impermanence and encourages the wise and virtuous use of our time on Earth.

Through the acceptance of death as part of the natural order, Stoics believe we can cultivate greater wisdom and serenity in life. This perspective helps focus our attention on what truly matters, enabling us to lead a virtuous and meaningful life unburdened by the fear of mortality.

Buddhist Perspectives

In Buddhism, death is viewed as part of the cycle of birth,

death, and rebirth, known as *samsara*. It is not the end but rather a transition to a new existence. The concepts of karma and reincarnation shape Buddhist views on death and decomposition, emphasizing the significance of moral conduct during life.

From a Buddhist perspective, death is a phase within the endless cycle of samsara. This cycle, governed by karma—the consequences of one's actions—determines the nature of an individual's rebirth and the happiness or suffering they will encounter in their next life. Karma underscores the interconnectedness of actions, life, and death.

Death in Buddhism is often seen as an opportunity for liberation from samsara. Through moral living, the cultivation of wisdom, and the practice of compassion, individuals can improve their karma and move closer to achieving *nirvana*, a state of ultimate freedom from suffering and rebirth.

This perspective encourages Buddhists to live mindfully and ethically. Since death is but a transition, it is not feared but accepted as a natural part of existence. Buddhist practice emphasizes the impermanence of all things and the importance of living fully in the present moment, free from attachment and fear.

By examining these various philosophical perspectives, it becomes clear that death is far more than a biological event. It is a profound philosophical and cultural phenomenon that shapes our understanding of existence and life itself. Engaging with these diverse views allows for a deeper comprehension of our mortality and offers multiple frameworks for approaching and understanding death.

8.1.2. The Concept of Immortality

Immortality of the Soul

Many religious and philosophical traditions hold the belief that the soul or consciousness persists beyond physical death.

These concepts are closely tied to notions of an afterlife or a spiritual existence.

The idea of the immortality of the soul appears across numerous religious doctrines and spiritual philosophies around the world. In Christianity, for instance, it is believed that the soul transitions to heaven, purgatory, or hell after death, depending on the individual's actions and faith during their lifetime. This belief in eternal life provides solace and hope, offering the promise of continued existence in a transcendent realm.

Similarly, in Hinduism, the concept of reincarnation plays a central role. It is believed that the soul (*Atman*) is reborn into a new body after death, continuing this cycle until it achieves *Moksha*—liberation from the cycle of birth and rebirth. This ongoing journey of the soul through various existences is governed by *karma*, which determines the quality of future lives based on past actions.

Greek philosophy also presents ideas about the immortality of the soul. Plato, for example, argued that the soul is eternal, existing before birth in the realm of ideas and returning to this immutable and timeless sphere after death. For Plato, the soul's immortality was evidence of a higher, spiritual reality beyond the physical world.

These diverse traditions and philosophies share the conviction that physical death does not signify the end of existence. Instead, the soul is viewed as an eternal and indestructible entity that transitions to another form or dimension. This belief in immortality provides comfort and shapes moral and ethical decisions in life, as people act with the understanding that their deeds have consequences in an afterlife or next existence.

Materialistic Perspectives
Materialist philosophers reject the notion of an immortal soul,

viewing consciousness as a product of physical processes in the brain. According to this perspective, death marks the cessation of consciousness, and the decomposition of the body is the definitive end of personal existence.

Materialism, a philosophical approach focused on the physical and measurable world, posits that consciousness is not an independent, immaterial entity but rather arises from the complex processes and structures of the brain. Thinkers like neuroscientist Antonio Damasio argue that thoughts, emotions, and consciousness itself can be fully explained by neural activity.

If consciousness is a product of brain function, it necessarily ends when the brain ceases to operate. The death of the physical body and its subsequent decomposition signify the absolute termination of individual existence. From this perspective, there is no continuation after death, no immortal soul, and no otherworldly realm.

For many materialists, however, this view does not lead to a negative or nihilistic outlook. On the contrary, the finite nature of life can serve as a call to live the present moment intensely and mindfully. Without the prospect of an afterlife, every moment and every decision in this life becomes more significant. This perspective encourages people to focus on creating a meaningful and fulfilling life in the here and now, as this is the only existence they will experience.

This materialistic worldview also influences ethical and moral beliefs. Without assuming an afterlife, moral decisions are grounded in the well-being and happiness of the living rather than potential consequences in a metaphysical beyond. This approach fosters an ethics centered on present human flourishing, rather than speculative considerations of a spiritual realm.

Summary

In conclusion, these two contrasting concepts—immortality of the soul and the materialistic perspective—highlight the diversity of philosophical reflections on death. While one viewpoint offers solace and hope through the idea of continued existence, the other emphasizes the importance of the present life and physical reality. Both perspectives invite us to reflect on the nature of consciousness and the meaning of life, providing distinct pathways for grappling with our own mortality.

8.2. Cultural considerations and rituals

8.2.1. Funeral Rites and Traditions

Christian Funeral Rites

In Christian traditions, death is often perceived as a transition to eternal life, and funeral rites are designed to reflect this belief. These rituals aim to honor the deceased, guide their soul toward heaven, and offer solace to grieving family members and friends. Ceremonies such as burial services, memorial gatherings, and prayers are central to the Christian approach to coping with death.

Christian funeral rites typically begin with a service held in a church or chapel. This service often includes hymns, scripture readings, prayers, and a sermon that reflects on the life and legacy of the deceased. The sermon frequently serves a dual purpose: it celebrates the individual's life while reinforcing the hope of resurrection and eternal life, as promised in Christian theology. The mourners find comfort in the shared belief that death is not an end but a gateway to a more profound spiritual existence.

Following the service, a burial ceremony is conducted. The coffin is carried to the gravesite, often accompanied by solemn hymns and psalms. At the grave, a final prayer is offered, commending the deceased's soul to God's care. This moment allows the mourners to bid a symbolic farewell while affirming their faith in an eventual reunion in the afterlife. The rituals surrounding a Christian funeral not only reflect deep religious convictions but also provide an opportunity for communal mourning and mutual support. The belief in eternal life and heavenly reunion offers immense comfort to those left behind.

Hindu Funeral Practices

In Hinduism, funeral practices are shaped by the belief in reincarnation and the cyclical nature of life and death. The cremation of the body is a central ritual, symbolizing the release of the soul (*Atman*) from the physical vessel and its journey toward rebirth or liberation. The ashes of the deceased are often scattered in a sacred river, such as the Ganges, to facilitate spiritual purification and the breaking of the cycle of rebirth.

The process begins with a purification ritual, during which the body of the deceased is cleansed and wrapped in white cloth, a symbol of purity and detachment. In some cases, the body is adorned with flowers, sandalwood paste, and other offerings that signify respect and sanctity. The family then gathers to pay their last respects, often performing rites intended to honor the deceased and invoke blessings for their soul.

The cremation ceremony takes place on a pyre, traditionally lit by the eldest son or a close male relative. Throughout the ritual, sacred mantras and prayers are recited to guide the soul on its spiritual journey. The act of cremation is believed to return the body's elements to the cosmos while freeing the soul for its next phase of existence. The scattering of ashes in a holy river is an essential part of the process, symbolizing the dissolution of physical bonds and the soul's unification with the divine.

Hindu funeral rites emphasize the interconnectedness of life, death, and spiritual liberation. They are not only a means of saying farewell but also an affirmation of the eternal nature of the soul and the importance of living a righteous life.

Buddhist Funeral Rituals

In Buddhism, funeral practices reflect the core teachings on impermanence, suffering, and the cyclical nature of existence (*Samsara*). Death is regarded as a natural transition rather than an endpoint, and the rituals are designed to support the deceased's journey toward a favorable rebirth or, ultimately, liberation (*Nirvana*).

Buddhist funerals often begin with the recitation of sutras, sacred texts that convey the Buddha's wisdom and provide guidance for the soul. Monks, family members, and friends gather to chant these texts, reflecting on the transient nature of life and the merits accumulated by the deceased. The recitation of sutras serves both to honor the deceased and to create a peaceful spiritual environment that facilitates their transition.

Meditation plays a crucial role in Buddhist funeral practices. Through meditation, participants focus their minds on compassion, understanding, and the impermanence of all things. This practice not only supports the deceased but also helps the bereaved to accept the reality of death and find inner peace. Offerings such as incense, flowers, and food are also common, symbolizing respect for the deceased and a connection to the spiritual realm.

In Tibetan Buddhism, additional rituals may span several days to guide the deceased through the intermediate state of existence (*Bardo*). These elaborate ceremonies aim to help the soul navigate the complexities of this transitional phase and secure a favorable rebirth. Tibetan Buddhist texts, such as the *Bardo Thodol* (commonly known as the Tibetan Book of the Dead), offer detailed guidance on how to assist the deceased during this critical time.

Conclusion

The diversity of funeral rites across Christian, Hindu, and Buddhist traditions reflects the profound cultural and spiritual significance of death. These rituals serve as a bridge between the living and the deceased, offering a means to honor the departed, seek closure, and find comfort in shared beliefs. Whether through prayers for eternal life, acts of purification and release, or meditative support for spiritual transition, each tradition provides a framework for coping with the universal human experience of mortality. These practices underscore the enduring importance of community, faith, and meaning-making in the face of life's inevitable end.

8.2.2. Cultural Perspectives on Decomposition

Western Views

In many Western cultures, death is often approached from a medical and hygienic perspective. Modern burial practices, such as embalming and cremation, aim to control the decomposition process, minimize its environmental impact, and delay the visible signs of bodily decay. These methods reflect a desire to manage death in a way that aligns with cultural aesthetics and sanitary standards.

Embalming, a common practice in Western funeral traditions, involves the injection of chemical preservatives into the body to slow decomposition and maintain a lifelike appearance. This process allows for extended viewings and public memorials, offering families the chance to say goodbye in a dignified and comforting manner. Beyond the aesthetic benefits, embalming fulfills practical needs, particularly when there is a delay between death and burial or transportation of the body over long distances.

Cremation is another widespread method, chosen for its practicality and symbolic resonance. In this process, the body is reduced to ashes through high heat. Cremation eliminates the need for long-term physical burial and prevents the natural process of decomposition from occurring in the ground. Families often scatter the ashes in meaningful locations, keep them in urns, or incorporate them into keepsakes, creating a lasting memorial to their loved ones.

Both practices reflect a broader Western tendency to distance death from its natural processes. Decomposition, with its raw physicality, is often perceived as unpleasant or undesirable, leading to efforts to sanitize and aestheticize the experience of death. The professionalization of funeral services further reinforces this approach, offering standardized and controlled environments for mourning that align with societal norms.

Traditional Cultures

In contrast, many traditional cultures embrace decomposition as a natural and integral part of the life cycle. These societies often incorporate rituals and practices that respect the natural process of decomposition while honoring the transition of the deceased into the next stage of existence.

In indigenous communities around the world, death is frequently seen as a continuation of life, not its end. For example, in some Native American traditions, burial rituals involve placing the body in close contact with the earth to facilitate its natural return to the soil. These practices emphasize the interconnectedness of life and death, viewing decomposition as a way for the deceased to give back to the earth and sustain future life.

The Zoroastrian tradition offers a unique perspective on decomposition. Followers of Zoroastrianism traditionally place their dead in *Dakhmas*, or "Towers of Silence," where the bodies are exposed to scavenging birds, such as vultures. This practice reflects a deep respect for the purity of the elements—earth, fire, and water—and a belief in the natural recycling of matter. By allowing nature to take its course, Zoroastrians see the body's decay as a meaningful contribution to the ecosystem.

Similarly, in Tibetan Buddhist sky burials, the body is dismembered and offered to birds of prey. This ritual symbolizes the impermanence of life and the interconnectedness of all beings. The act of providing sustenance to other creatures is seen as a final act of generosity and a way to dissolve attachments to the physical self.

These traditional practices reflect a profound acceptance of the natural order. Death and decomposition are viewed not as events to be feared or controlled but as essential transitions within the broader cycles of nature. Rituals accompanying these processes often aim to honor the deceased and celebrate their integration into the earth's ongoing life systems.

Conclusion

Cultural perspectives on decomposition reveal a spectrum of beliefs and practices, ranging from controlled, sanitized approaches in Western societies to integrative, nature-focused traditions in many indigenous and traditional cultures. Western methods, such as embalming and cremation, often aim to delay or alter the visible signs of decomposition, reflecting a preference for maintaining control and distance from the raw realities of death. In contrast, traditional practices highlight a harmonious relationship with nature, seeing decomposition as a sacred and necessary process.

These differing views illustrate how cultural, philosophical, and religious values shape the ways societies confront death and honor the dead. By exploring these diverse approaches, we gain a deeper understanding of the universal human experience of mortality and the rich tapestry of practices that seek to find meaning and continuity in the face of life's inevitable end.

8.2.3. Symbolism of Death and Decomposition
Symbolic Significance
In many cultures, death and decomposition hold profound symbolic meanings, often representing transformation, renewal, and the cyclical nature of life. These themes are frequently explored in art and literature, serving as powerful reminders of life's impermanence and the continuity of natural cycles.

Across various artistic traditions, death has been a central motif used to confront the fragility of human existence. In Western art, for example, the vanitas still-life paintings of the 17th century are emblematic of this theme. These works often feature symbolic objects such as skulls, wilting flowers, and hourglasses, all designed to evoke the transitory nature of life and the inevitability of death. These visual metaphors serve as moral reminders to live meaningfully and reflect on the fleeting nature of earthly achievements.

In literature, death and decomposition frequently symbolize deeper existential truths. William Shakespeare's *Hamlet* provides a poignant exploration of these themes. The protagonist's famous "To be or not to be" soliloquy delves into the uncertainty and inevitability of death, reflecting on its role as a natural and inescapable aspect of human existence. Elsewhere in the play, Hamlet's contemplation of Yorick's skull underscores the equality and inevitability of mortality, framing decomposition as a unifying force that transcends social and personal distinctions.

In indigenous cultures, death and decomposition are often seen as integral to the natural order, emphasizing the cycle of life, death, and rebirth. Among Native American traditions, for instance, death is frequently understood as a return to the earth and the Creator, signifying renewal and continuity within a greater cosmic cycle. These beliefs underscore the interconnectedness of all living things and the idea that decomposition nourishes new life, perpetuating the balance of nature.

Ritualized Decomposition

Some cultures have developed specific rituals that symbolically or practically honor the process of decomposition. These practices often reflect profound spiritual and ecological values, connecting the act of decay to broader ideas of transformation and renewal.

One striking example is the Tibetan practice of *sky burials*. In this ritual, the bodies of the deceased are placed in high, remote locations and offered to scavenging birds such as vultures. Rooted in Buddhist beliefs about impermanence, this act acknowledges the transience of the physical form while focusing on the liberation of the spirit. Feeding the body to birds is seen as a final act of generosity, symbolizing the deceased's contribution to the continuation of life. This deeply symbolic practice reinforces the idea that death is not an end but a vital part of life's ongoing cycle.

In the West, *tree burials* have gained popularity as a modern interpretation of decomposition's transformative power. This practice involves placing the ashes of the deceased in biodegradable urns that contain seeds, allowing a tree to grow in their memory. The growing tree serves as a living monument to the deceased, symbolizing renewal, life's continuity, and the return to nature. This environmentally friendly approach not only honors the deceased but also fosters ecological restoration, creating a lasting, positive impact on the environment.

Conclusion

The symbolism of death and decomposition is deeply embedded in cultural rituals and artistic expressions across the world. While Western traditions often focus on controlling and aestheticizing death, many indigenous and traditional practices embrace decomposition as a sacred process integral to life's cycle. Rituals like sky burials and tree burials highlight the universal human desire to find meaning in death and ensure that it contributes to the ongoing cycle of life.

By examining these diverse cultural perspectives, we gain insight into how humanity has sought to understand and honor the inevitability of death. Whether through art, literature, or ritual, these practices reflect the profound human capacity to find beauty and significance in the processes of transformation and renewal that define the natural world.

Energy and Transience -The Journey After Death

8.3. Decomposition's Role in Philosophy and Ethics

8.3.1. Environmental Ethics and Sustainability

Sustainable Burial Practices

Contemporary philosophy increasingly explores the ethical dimensions of burial practices and their environmental impacts. A growing movement advocates for sustainable burial methods, such as ecological funerals and biodegradable burial options, designed to minimize the ecological footprint associated with death.

Modern philosophical discourse intersects with environmental ethics to address the broader implications of burial practices on the planet. With global environmental concerns escalating, philosophers and environmentalists alike have sought ways to reframe death as an opportunity to promote ecological responsibility. This has given rise to the concept of *green burials*, which aim to harmonize burial practices with the natural world while reducing harmful impacts.

Green burials encompass a variety of environmentally conscious approaches. Biodegradable caskets and burial shrouds are increasingly popular, designed to facilitate the natural decomposition of the body without releasing toxic substances into the soil. These materials allow the deceased to return to the earth in a way that supports the ecosystem rather than burdening it. Additionally, natural burials often involve interring the body in conservation or wilderness areas, forgoing headstones or elaborate grave markers to preserve the natural landscape.

Another innovative approach is the composting of human remains, also known as *recomposition*. This process involves transforming the body into nutrient-rich soil under controlled conditions. Emerging primarily in the United States, this method represents a groundbreaking alternative to conventional burial and cremation, significantly reducing carbon emissions and resource use. The resulting compost can be used to enrich soils, directly contributing to the cycle of life and embodying the concept of "from dust to dust" in an ecologically meaningful way.

Ethical Considerations

Ethical discussions surrounding decomposition and the treatment of the body after death encompass concerns of environmental justice and respect for the deceased. These considerations are pivotal in shaping modern burial practices and policies for body disposal.

Respect for the deceased is a core value in many cultures, with rituals and practices designed to honor the body and the person it represents. Modern sustainable practices must balance this respect with ecological responsibility, ensuring that the dignity of the deceased is preserved while minimizing environmental harm. This balance often involves integrating traditional values with innovative practices, fostering a deeper appreciation for the interconnectedness of human life and nature.

Environmental justice is another significant aspect of these ethical considerations. Access to eco-friendly burial options is often limited by socioeconomic factors, with marginalized communities disproportionately affected by the environmental impacts of conventional practices. For example, cremation, while perceived as environmentally friendly, contributes to air pollution and carbon emissions, and traditional burials can involve chemicals that leach into the soil. Ensuring equitable access to sustainable burial methods is an ethical imperative, addressing disparities in how communities experience the ecological and economic consequences of death.

Transparency within the funeral industry also plays a vital role in promoting ethical practices. Many individuals remain unaware of the environmental implications of different burial methods, which can result in uninformed decision-making. Public education and awareness initiatives are essential for empowering individuals and families to choose options that align with both their values and ecological priorities.

Conclusion

The modern discourse surrounding decomposition and burial practices reflects a growing sensitivity to environmental and ethical responsibilities. By reimagining the way we handle death, society can move toward practices that honor both the deceased and the planet. Sustainable burial methods, such as green burials and recomposition, not only reduce ecological impacts but also provide meaningful ways for individuals to contribute to the preservation of the environment, even in death.

This evolving ethical framework emphasizes the need for burial practices that are not only environmentally sustainable but also culturally respectful and accessible to all. As these discussions advance, they hold the potential to transform our relationship with death, fostering a deeper awareness of its role in the cycles of life and its capacity to inspire ecological stewardship.

8.3.2. The Role of Decomposition in Cultural Identity

Cultural Identity

The ways in which cultures approach death and decomposition significantly shape their cultural identity and sense of community. Burial rituals and traditions are deeply embedded in cultural values and beliefs, serving as vital mechanisms for fostering cultural cohesion and preserving collective memory.

In many societies, burial practices and attitudes toward the decomposition of the body are intrinsic to the cultural framework. These practices reflect a community's core values and beliefs about life, death, and the afterlife, shaping both individual and collective identities. Rituals surrounding death are often designed to honor the deceased, provide solace to the living, and maintain a sense of unity within the community.

For instance, elaborate burial ceremonies in some cultures involve intricate rites, prayers, and symbolic actions that emphasize the importance of the departed individual and their continuing influence on the living. Such ceremonies often serve to reinforce social bonds, offering the bereaved an opportunity to collectively grieve and celebrate the life of the deceased. These rituals not only help individuals process their loss but also strengthen the cultural fabric by affirming shared values and traditions.

One prominent example is Mexico's **Día de los Muertos (Day of the Dead)**, where deceased loved ones are remembered and celebrated annually. This tradition, rooted in Indigenous and Catholic practices, brings communities together through vibrant altars (ofrendas), food offerings, and symbolic decorations. It illustrates how death and decomposition can be integrated into cultural identity as part of a broader cycle of life and renewal. Such rituals ensure continuity between past and present, maintaining intergenerational connections and fostering a profound sense of belonging.

Global Perspectives

Globalization has facilitated an exchange of funeral practices and traditions, often blending cultural elements and fostering new perspectives on death and decomposition. This cross-cultural interaction is reshaping how communities navigate the evolving landscape of death rituals while embracing the diversity of human experiences related to mortality.

As cultures become more interconnected, funeral practices from one region often influence those in another, creating opportunities for the emergence of hybrid traditions. For example, the growing acceptance of **sustainable burial practices**—initially popularized in Western countries—is now being adopted in regions with different traditional customs, such as Japan and South Korea. These eco-friendly approaches, which include biodegradable caskets and natural burials, are being adapted to align with local cultural values, resulting in rituals that honor both environmental concerns and traditional beliefs.

This blending of practices reflects a dynamic interplay between tradition and innovation. For instance, sustainable burial methods in regions with strong ancestral worship traditions often incorporate rituals that honor the deceased while adhering to environmentally conscious principles. This integration fosters cultural evolution while preserving respect for longstanding customs.

Moreover, globalization allows for greater access to diverse perspectives on death, enabling individuals and communities to reimagine their own traditions. Exposure to different cultural practices through media, travel, and academic discourse broadens understanding and inspires the adoption or adaptation of new rituals. This global exchange encourages greater appreciation for the myriad ways humanity approaches death and decomposition, creating opportunities for mutual respect and learning.

Conclusion

The ways in which cultures handle death and decomposition are deeply tied to their identity, values, and collective memory. Burial practices, whether rooted in tradition or influenced by global interactions, reflect the unique perspectives of each culture while highlighting shared human concerns about mortality and remembrance.

In an increasingly interconnected world, these practices are evolving, blending traditional values with modern approaches to create meaningful rituals that resonate across diverse cultural contexts. By embracing both local and global perspectives, communities can honor their heritage while addressing contemporary challenges, fostering an inclusive and dynamic understanding of life, death, and the cycles that connect them.

8.4. Summary

Philosophical and cultural perspectives on death and decomposition reflect a diverse range of beliefs and traditions deeply embedded in human history and thought. These perspectives significantly shape how societies confront death and integrate the concept of decomposition within their cultural frameworks.

Exploring these aspects provides valuable insights into human existence, cultural identity, and the ethical considerations surrounding sustainable practices. By understanding various philosophical approaches to death—such as existentialist, stoic, and Buddhist perspectives—we can appreciate the profound meanings these topics hold across different worldviews. Each philosophy offers unique insights that influence how individuals and societies perceive and process death and decomposition.

Existentialist philosophers like Jean-Paul Sartre and Albert Camus emphasize death as a central element of human existence, serving as a catalyst for meaning and authenticity. Stoic thinkers such as Epictetus and Seneca advocate for accepting death as a path to inner peace and wisdom. Meanwhile, Buddhist views frame death as part of the cyclical process of birth, death, and rebirth, underscoring the ethical importance of one's actions in life.

Cultural considerations and rituals surrounding death are equally intricate. Burial practices such as Christian funerals, Hindu cremations, and Buddhist ceremonies reflect deeply rooted cultural values. These rituals not only foster cultural cohesion but also strengthen collective memory. In many cultures, death symbolizes not just the end of life but also transformation and renewal, a theme echoed in symbolic and ritualized decomposition processes.

Contemporary philosophy and ethics increasingly focus on the ecological and ethical implications of burial practices. Sustainable burial methods, such as green burials and biodegradable options, minimize ecological footprints and promote environmental justice. These approaches ensure that burial practices are not only respectful toward the deceased but also environmentally conscious and sustainable.

Additionally, globalization highlights how burial practices and traditions are exchanged and blended across cultures, offering new perspectives on managing death and decomposition. This cultural exchange fosters deeper understanding and greater appreciation of the diverse ways humanity experiences mortality. It also facilitates the development of meaningful new rituals that resonate within evolving cultural landscapes.

In summary, philosophical and cultural reflections on death and decomposition underscore their importance in shaping individual and collective experiences. By integrating ethical, ecological, and cultural considerations, societies can honor their traditions while addressing contemporary challenges, creating practices that are both meaningful and sustainable.

Conclusion

Summary of Key Insights
This book has thoroughly explored how energy in the human body transforms after death, revealing profound implications for both the environment and ecological systems. Detailed

analyses were provided on the processes and stages that occur from the moment of death through decomposition to long-term energy transformations. It became clear that an individual's death is not an absolute end but part of a broader, continuous natural cycle. By examining these phenomena, we have gained a deeper understanding of the energetic transformations that follow death and their integration into the larger web of life.

The Human Body as an Energy System

The human body operates as a complex energy system where chemical, electrical, and mechanical energies are constantly converted. These energy changes persist after death, as the body transitions into decomposition through chemical processes. The mechanisms of these transformations significantly impact subsequent decomposition and the return of nutrients and energy to the ecosystem.

During life, these processes sustain bodily functions, but after death, they shift in nature and purpose. The chemical energy once utilized to sustain life is slowly released and transformed to support natural cycles. The human body, once a self-contained entity, becomes a part of the broader ecological system, with every component contributing to the sustenance of life.

The Final Moments of Life

The last moments before death are marked by a series of physiological changes, including the cessation of heartbeat and respiration. This halts blood flow and reduces the nutrient supply to tissues, triggering biochemical reactions that facilitate the transition from a living system to a non-living organism.

In this critical phase, the heart stops pumping oxygen and nutrients to cells, leading to an immediate cessation of circulation. This initiates a cascade of biochemical changes that mark the definitive departure of life.

At the Moment of Death

Death sets off a series of biochemical and physiological changes that lead to the body's decomposition. These initial processes include the breakdown of adenosine triphosphate (ATP) into adenosine diphosphate (ADP) and adenosine monophosphate (AMP), facilitated by enzymes and bacteria. These changes initiate the early stages of decomposition, which rapidly propagate throughout the body.

Death marks the start of a complex disintegration process, transforming the once-living body into its fundamental components. This process is biologically fascinating and ecologically significant, as it facilitates the recycling of energy and nutrients into the environment.

The First Hours After Death

In the hours following death, rigor mortis (stiffening of muscles) and autolysis (self-digestion) set in. These processes involve enzymatic transformations within cells and the onset of bacterial decomposition. Changes in body temperature and chemical composition influence the pace and progression of decomposition.

Rigor mortis, caused by chemical changes in muscles, results in temporary stiffness of the body. Simultaneously, autolysis begins, as cells release their enzymes to break down internal structures. These processes are pivotal in initiating decomposition and setting the stage for subsequent transformations.

The Process of Decomposition

Decomposition is a multi-stage process, progressing from microbial and insect activity to the complete mineralization of organic material. The phases of decomposition—from early decay to skeletonization—illustrate the ongoing transformation of organic energy into inorganic substances, returning nutrients to the soil and ecosystem.

Microorganisms and insects play a central role in this breakdown, disassembling complex molecules into simpler substances. This gradual release of stored energy and recycling of nutrients supports the environment and reflects a natural sequence of biological activities, ultimately integrating the body into the ecological cycle.

Long-Term Energy Transformations

Long-term energy transformations involve converting organic substances into minerals and integrating these nutrients into the soil's nutrient cycle. The formation of humus and changes in soil structure and chemistry are crucial for soil fertility and ecological balance.

These processes sustain soil quality and ecosystem productivity. The conversion of organic materials into humus releases valuable nutrients that promote plant growth and other life forms. The enhancement of soil structure through humus formation helps preserve soil fertility and fosters ecological health.

Ecological and Physical Perspectives

Decomposition has far-reaching effects on ecological systems and physical processes. These include influencing nutrient cycles, altering soil biology and structure, and causing physical changes like temperature and moisture variations. Understanding these processes is central to environmental functions and sustainable resource use.

Decomposition regulates soil moisture and temperature, affecting microclimates and living conditions for numerous organisms. These ecological interactions reveal the interconnectedness of life and death processes and their collective impact on ecosystem functionality.

Philosophical and Cultural Considerations

Death and decomposition are deeply interwoven with

philosophical and cultural reflections. Different cultures and philosophies present diverse views on death, decomposition, and the afterlife, which significantly shape rituals, ethical considerations, and cultural identity in dealing with mortality. The way a society perceives and handles death reflects its core values and beliefs.

In many cultures, death is seen as a transition and transformation, which is expressed through burial rituals and practices of remembrance. Philosophical reflections on death and decomposition delve into the deeper meanings of these processes and their role within the context of life. Cultural rituals and traditions provide a framework for collective mourning and remembrance, contributing to the preservation of cultural identity.

The diversity of perspectives on death highlights both the universal and unique aspects of human experiences and beliefs regarding the end of life. Through these practices and reflections, societies maintain a connection between the living and the deceased, while also affirming their cultural and spiritual heritage.

Broader Perspective on Energy Transformations

By examining the processes and stages of death and decomposition, this book has offered a comprehensive perspective on their ecological, physical, and cultural implications. It underscores that death is not merely the end of an individual's life but an essential part of the natural cycle that sustains and nurtures life on Earth. Understanding these processes helps us appreciate the interconnectedness of nature and the role of death in maintaining ecological balance.

Implications and Outlook

Impacts on Environmental and Sustainability Practices

Understanding decomposition and the associated energy

transformations can contribute to the development of sustainable burial practices and environmentally friendly methods. Incorporating ecological considerations into funerary customs, such as green burials and biodegradable products, can minimize the ecological footprint and support environmental conservation.

Ecological burial methods, such as the use of biodegradable coffins and urns, promote the return of nutrients to the soil, facilitating the natural regeneration of ecosystems. These practices not only reduce environmental impact but also support sustainability and biodiversity. Additionally, integrating these ecological principles into funerary traditions encourages a more conscious approach to natural resources and raises awareness about sustainability across all facets of life.

Influence on Science and Research

Research into decomposition and energy transformations provides valuable insights into biochemistry, ecology, and environmental science. These findings can improve models of nutrient cycles, inform the development of waste management technologies, and deepen understanding of biological processes.

Scientists can use these insights to develop more precise models of decomposition, capturing the interactions between microorganisms, insects, and chemical reactions. Such models can contribute to creating efficient waste management strategies that mimic natural decomposition processes, thereby reducing environmental harm.

Moreover, advances in biochemistry and ecology derived from these studies may lead to innovative approaches to enhancing soil fertility and restoring degraded ecosystems. Overall, this research expands our understanding of the natural world and opens new possibilities for addressing environmental challenges through science and technology.

Ethics and Culture

Ethical considerations surrounding the treatment of human remains and decomposition are central to fostering respectful and culturally sensitive funerary practices. Reflecting on cultural and philosophical perspectives can deepen our understanding of human existence and the diversity of approaches to death and decomposition.

Different cultures maintain rituals and traditions deeply rooted in their philosophical and religious beliefs. Recognizing and valuing this diversity can help develop practices that are respectful and empathetic to the needs and values of various communities. Ethics play a pivotal role in shaping funerary practices that respect the deceased while also considering environmental impacts.

By engaging with these ethical questions, we can encourage practices that uphold individual dignity and protect ecological integrity, creating a balance between honoring cultural traditions and fostering environmental sustainability.

Concluding Thoughts

Death and decomposition are central aspects of the life cycle, deeply rooted in natural and cultural processes. Through the comprehensive exploration of these topics, we have gained insight into the complex interactions among biological, ecological, physical, philosophical, and cultural dimensions. These findings enhance our understanding of nature and our own existence, offering valuable perspectives for practice and research. Engaging with the processes of death and decomposition not only fosters a deeper appreciation of the natural world but also encourages a respectful and reflective attitude toward life and the environment.

It is our responsibility to use this knowledge to identify sustainable and ethical pathways that align with the natural world while respecting cultural diversity. By recognizing the natural processes of death and decomposition as integral parts of life, we can develop practices that are both ecologically sustainable and culturally sensitive. This enables us to adopt a holistic approach that maintains a harmonious balance between the natural world and human culture. Ultimately, this opens the possibility of shaping our lives and society in ways that protect the environment while preserving human dignity and cultural identity.

Through this holistic perspective, we can contribute to a sustainable and ethically responsible future that respects and preserves the legacy of both nature and humanity.

Energy and Transience - The Journey After Death

Appendix

A.1 Glossary of Terms

ATP (Adenosine Triphosphate)
ATP is a high-energy molecule found in all living cells and serves as the primary energy source for many biochemical processes. It is produced through cellular respiration and, upon death, is converted into ADP (Adenosine Diphosphate) and AMP (Adenosine Monophosphate).

Autolysis
This refers to the process by which enzymes produced by the cells break down the cells' own structure, contributing to the decomposition of the body after death.

Bacterial Decomposition
Bacterial decomposition involves the breakdown of organic material by microbes, such as bacteria, that are either already present in the body or enter from the external environment. This process is a crucial component of decomposition.

Humus
Humus is the organic matter found in the soil that is formed from the decomposition of plant and animal remains. It plays a vital role in improving soil structure and fertility.

Carbon Cycle
The carbon cycle is the biogeochemical cycle through which carbon exists in various forms (such as carbon dioxide, organic matter, and carbohydrates) as it circulates among the atmosphere, biosphere, and soil.

Nutrient Cycle
This process describes how nutrients like nitrogen, phosphorus, and potassium circulate through the environment, beginning with their uptake by plants, followed by consumption by animals, and ultimately their return to the soil through decomposition.

Rigor Mortis

Rigor mortis refers to the stiffness of the muscles that occurs after death due to chemical changes within the muscle cells. This physiological phenomenon marks a significant stage in the process of decomposition.

A.2 References

1. ***Harris, J. (2021)***. *Biological Processes of Decomposition*. Springer Nature.
   - This book provides a comprehensive overview of the biological processes of decomposition and the role of microbes in these processes.

2. ***Smith, R. (2019)***. *Ecological Impact of Decomposition*. Oxford University Press.
   - A detailed analysis of the ecological impacts of decomposition on nutrient cycles and soil biology.

3. ***Williams, M. & Clark, T.*** (2020). *Philosophy and Death: Perspectives and Interpretations*. Routledge.
   - A philosophical examination of various perspectives on death and decomposition, including existentialist and stoic viewpoints.

4. ***Davis, L. (2022)***. *Cultural Rituals and the Afterlife*. University of Chicago Press.
   - A book about various cultural and religious rituals related to death and decomposition around the world.

5. ***Miller, S. & Anderson, B. (2021)***. *Sustainable Burial Practices*. Harvard University Press.
   - An examination of sustainable burial practices and their impact on the environment.

A.3 Methodology

This book is grounded in a combination of literature review, scientific studies, and philosophical texts. The primary methods employed include:

Literature Review:

A comprehensive examination of academic literature and scientific articles related to the topics of decomposition, energy transformations, ecological impacts, and cultural perspectives.

Case Studies:

An analysis of case studies concerning decomposition in various environments and under different conditions.

Philosophical Analysis:

An exploration and interpretation of philosophical texts and theories pertaining to death, decomposition, and life after death.

A.4 Sources for Further Information

Online Databases and Scientific Journals:

- Google Scholar
- PubMed
- JSTOR

Organizations and Institutions:

- American Society of Microbiology (ASM)
- Soil Science Society of America (SSSA)
- International Cemetery, Cremation and Funeral Association (ICCFA)

Websites:

- Environmental Protection Agency (EPA) – Information on sustainable burial practices.
- The Center for the Study of Death and Society

A.5 Acknowledgments

We would like to extend our gratitude to all the researchers, authors, and institutions whose works and studies provided the foundation for the content of this book. Special thanks go to the experts in the fields of biology, ecology, philosophy, and cultural studies for their valuable contributions and support. Their work has helped develop a deeper understanding of the complex processes of death and decomposition and has illuminated the connection between scientific research and cultural practices.

A.6 Liability

The information contained in this book is intended for general informational purposes only and does not constitute legal or medical advice.

A.7 Contact Information

Email: chai2023@gmx.net
Discord: https://discord.gg/QMt4DBGr
Facebook: Drake Graeve

www.ingramcontent.com/pod-product-compliance
Lightning Source LLC
Chambersburg PA
CBHW052245220526
45471CB00001B/203